An Introduction to

WATER QUALITY IN RIVERS, COASTAL WATERS AND ESTUARIES

by Desmond Hammerton
OBE, BSc, CBiol, FCIWEM, FIMgt, FRSE

The Chartered Institution of Water and Environmental Management

©The Chartered Institution of Water and
Environmental Management, 1996
ISBN 1 870752 27 9

The Chartered Institution of Water and Environmental Management
15 John Street, London WC1N 2EB

FOREWORD

The Chartered Institution of Water and Environmental Management (CIWEM) is a multi-disciplinary professional and examining organization for engineers, scientists and other professional personnel engaged in water and environmental management. It was formed in July 1987 by the unification of three eminent organizations, The Institution of Public Health Engineers, The Institution of Water Engineers and Scientists and the Institute of Water Pollution Control, each having a history of some 100 years.

Over the years the predecessor bodies have produced definitive manuals and other publications, notably in respect of British practice in the water industry. These have become reference sources for those who are actively engaged in the field, as well as for students seeking authoritative guidance in preparing for professional qualifications. Such publications are being continued by CIWEM, and the range is being extended to take account of the wider environmental issues and interests which the new organization now embraces.

This introductory book on *Water Quality in Rivers, Coastal Waters and Estuaries* is the fifth in a new series being published by CIWEM on an aspect of environmental management which is today giving rise to increasing concern. Although it is intended as an introduction to the subject, and has been written as a general guide for the interested layperson, the book provides a comprehensive summary of the situation in the UK, and should also be useful to those embarking on a career in the field.

The Institution wishes to record its thanks to those members who have contributed to the production of the book, and in particular to Professor Desmond Hammerton who has been responsible for the preparation of the text and to Ted Bruce the Editor.

January 1997   **Tricia Henton**
                                                                       President

ACKNOWLEDGEMENTS

Plates 1 to 10 are reproduced by courtesy of the former Clyde River Purification Board.
 The Institution is grateful to the Board, and to the many authors whose contributions, in the form of figures and tables, are acknowledged individually.

PREFACE

About forty years ago my father, then Chief Chemist of the Metropolitan Water Board, London, received a visit from a Scotland Yard detective who was seeking help in solving an unusual crime. A large consignment of whisky had arrived in Paris whereupon it was found that in every bottle the whisky had been replaced by water. The detective provided one of the bottles and asked if chemical analysis could suggest where the fraud had taken place. In fact chemical analysis showed that the dissolved constituents in the water corresponded exactly to those of Glasgow's water supply from Loch Katrine, one of the 'purest' water supplies in the UK.

This story illustrates the fact that water, however pure, contains a wide range of dissolved substances and that no two fresh waters from different sources are ever exactly alike but reflect the local environmental factors such as atmospheric conditions, geology of the river or lake catchment and the land use within the catchment. Even sea waters which are remarkably uniform in quality show distinct variations in composition from one location to another.

This book is intended to provide a guide to the subject of water quality in the natural environment* for those in the water industry and the public who are not professionally involved in its measurement or its management. It is designed to provide a simple but reasonably comprehensive introduction to the subject and should also be of value to those who may be contemplating a career involved with water quality.

During the preparation of this book, new regulatory authorities – the Environment Agencies – have come into being and have taken over the functions of the previous authorities which are detailed on page 55. I believe that this change will further strengthen and improve environmental management because account will now be taken of all environmental pathways. However, the principles of water quality management outlined in this book will, of course, remain unchanged.

I wish to thank my colleagues Dr Andrew Haig and Dr Ken Pugh of the former Clyde and North East River Purification Boards for their helpful comments and suggestions on the draft text. I am grateful to John Davis of the Water Research Centre, Medmen-

*An Introduction to Drinking Water Quality by Norman Nicholson is a companion volume in this series.

ham for assistance with certain technical data and particularly to Ian Currie of the Tweed River Purification Board who first suggested that I should write this book and for his encouragement throughout. Finally I owe a great debt of gratitude to my wife for her support and help and her invaluable comments on the clarity and readability of the text throughout its preparation.

DESMOND HAMMERTON

CONTENTS

	Page
Foreword	iii
Acknowledgements	iv
Preface	v
Principal Abbreviations Used in the Text	x
Introduction	1

1. **Characteristics of Natural Waters** 2
   The Water Cycle 2
   Definition of Water Quality 4
   Properties of Water 5
   Classification of Water Bodies 11

2. **The Measurement of Water Quality** 13
   Physical Characteristics 13
   Chemical Characteristics 18
   Biological Characteristics 30
   Impact of Pollutants 42
   Water Quality Assessment 50
   Quality Control 52

3. **Legislation and Management of Water Quality** 53
   UK and EC Legislation 53
   International Agreements 54
   Regulatory Authorities 55
   Consent System 55
   Integrated Pollution Control 57
   Statutory Water Quality Objectives 57
   Catchment Management Plans 58

4. **Environmental Quality Standards and Water Quality Classification** . . . 59
   Environmental Quality Standards 59
   Water Quality Classification 61
   Quality of UK Waters 63

5. Future Outlook 68
Appendix. European Community Legislation 81
Bibliography and Further Reading 82
Index 84

FIGURES

	Page
1. Water residence times in inland water bodies	4
2. Dependence of specific gravity of water and ice on temperature	6
3. Typical temperature profile of a stratified lake or reservoir	8
4. Relative proportions of free carbon dioxide, bicarbonate and carbonate in water in relation to pH	17
5. Diagram of a 'trophic pyramid'	34
6. Simplified food web for organisms in a mature river	35
7. Diatom diagram and pH reconstruction for Round Loch of Glenhead	39
8. Biological indicators of pollution	41
9. Some lake zooplankton	43
10. Recovery of Clyde Estuary	44
11. Diagrammatic presentation of effects of an organic effluent on a river	47
12. Bathing waters: compliance with coliform standard by region and country	67

PLATES

(between pages 71 and 80)

1. Gas chromatograph/mass spectrometer in use in laboratory
2. Use of electro-fishing apparatus to study fish populations
3. Collection of a sample of stream fauna by 'kick' sampling with a hand net
4. Sieving a grab sample of bottom sediment
5. Washing biological samples in a ventilated sink
6. Laboratory examination of biological samples
7. Stream sampling for chemical analysis
8. Survey vessel for estuaries and coastal waters
9. Sampling distillery effluent in inshore waters
10. Reversing water bottle for collection of deep samples in coastal waters

TABLES

	Page
1. The hydrological cycle	3
2. Typical concentrations of major ions in rainfall	19
3. Chemical composition of river water	21
4. Dissolved inorganic constituents of natural waters	22
5. Solubility of oxygen in pure water at sea level	28
6. Examples of Environmental Quality Standards	32
7. River water quality in England and Wales, 1958–80	63
8. River water quality in England and Wales, 1980–90	63
9. River water quality in Scotland, 1980–90	64
10. Estuary water quality in Scotland, 1990	65
11. Estuary water quality in England and Wales, 1990	65
12. Coastal water quality in Scotland, 1990	66

PRINCIPAL ABBREVIATIONS USED IN THE TEXT

ASPT	Average score per taxon
BOD	Biochemical oxygen demand
CMP	Catchment management plan
COPA II	Control of Pollution Act 1974, Part II
DO	Dissolved oxygen
DoE	Department of the Environment
EA	Environment Agency
EC	European Community
EQS	Environmental quality standard
GEMS	Global environmental monitoring system
HMIP	Her Majesty's Inspectorate of Pollution
HMIPI	Her Majesty's Industrial Pollution Inspectorate
HMS	Harmonized monitoring scheme
IPC	Integrated pollution control
NRA	National Rivers Authority
RCEP	Royal Commission on Environmental Pollution
RIVPACS	River invertebrate prediction and classification system
RPA	River Purification Authority
RPB	River Purification Board
SEPA	Scottish Environment Protection Agency
SOEnD	Scottish Office Environment Department
SS	Suspended solids
SWQO	Statutory water quality objective
TDS	Total dissolved solids
TOC	Total organic carbon
UNCED	United Nations Conference on Environment and Development
UNECE	United Nations Economic Commission for Europe
WHO	World Health Organization

INTRODUCTION

Water is the basis of life on earth and therefore our most important resource. It is generally accepted that life started in water and only later began to migrate to the land when organisms were sufficiently evolved to 'carry' their original environment with them in their cellular fluids and in the sap of higher plants and the blood of animals. A reliable supply of water is essential for the establishment of biological communities, and dry climatic regions such as deserts can only be populated by plants and animals which have developed special mechanisms for retaining water and replacing any losses. Our modern societies use ever-increasing amounts of water for domestic and industrial purposes, and this has led to problems of supply and quality on a global scale.

In recent years the quality of our water resources has become a major cause of public concern, not only because of its importance for human health (through the provision of safe drinking water) but also because of the impact of water pollution on freshwater and marine life. In order to control pollution and manage water quality it is essential to make use of classification systems which provide an accurate and reliable measure of water quality and, by regular analysis, indicate whether that quality is improving or deteriorating with the passage of time.

This volume is intended to provide a simple guide to present-day systems of water quality measurement in the context of UK and European Community legislation and international agreements. It is aimed at those in the water industry who are not directly involved in water quality measurement and also at a wider readership of all those who are interested in the quality of our aquatic environment and the environmental agencies whose job it is to maintain and improve that quality. For those who wish to study the subject in more detail a list of references and suggestions for further reading is included.

1. CHARACTERISTICS OF NATURAL WATERS

THE WATER CYCLE

In order to understand the quality characteristics of natural waters it is helpful to consider the water cycle which connects the various water bodies (Table I and Fig. 1). It is estimated that water comprises about 7% of the earth's mass and that 94% of the total water is contained in the oceans and seas and is highly saline. Of the freshwater component about one third is locked up in the icecaps and glaciers and two thirds in groundwaters while lakes and rivers represent only 0.14%.

Seawater, which covers about 75% of the earth's surface, is remarkably constant in chemical composition wherever it is sampled. The only exceptions to this are nearly closed seas such as the Baltic, which is brackish because of the large freshwater input, and the Red Sea which has a somewhat higher salinity resulting from high evaporation and no significant freshwater input.

This volume is concerned with water quality in rivers and lakes in its journey from upland sources to the sea and the impact of rivers and other direct discharges on coastal water quality. In contrast to the relatively uniform quality of seawater, freshwaters vary enormously in quality depending on climate, geology, soils, vegetation, land use and other factors within each river basin. Precipitation, whether in the form of rain, snow or ice (hail) washes gases and dust particles out of the atmosphere. The gases include oxygen, carbon dioxide, ammonia and, from industrial sources, sulphur dioxide, carbon monoxide, oxides of nitrogen and hydrocarbons. Particulates from natural sources include windblown soil particles and particles from volcanic eruptions while man-made particles include fuel ash and metallic substances.

Rainfall has always been slightly acidic due to the natural formation of carbonic acid, but has become much more acidic during this century due to the formation of

Table I. The Hydrological Cycle

	Total cycle volume (10^6 km^3)	Freshwater volume only (%)	Freshwater volume without icecaps and glaciers (%)	Residence times
Reservoirs				
Oceans and seas	1370	94			~4,000 years
Lakes and reservoirs	0.13	<0.01	0.14	0.21	~10 years
Swamps and marshes	<0.01	<0.01	<0.01	<0.01	1–10 years
River channels	<0.01	<0.01	<0.01	<0.01	~2 weeks
Soil moisture	0.07	<0.01	0.07	0.11	2 weeks–1 year
Groundwater	60	4	66.5	99.65	2 weeks–50,000 years
Icecaps and glaciers	30	2	33.3		10–1,000 years
Atmospheric water	0.01	<0.01	0.01	0.02	~10 days
Biospheric water	<0.01	<0.01	<0.01	<0.01	~1 week
Fluxes				
Evaporation from oceans	425			
Evaporation from land	71			
Precipitation from oceans	385			
Precipitation from land	111			
Run-off to oceans	37.4			
Glacial ice	2.5			

Source: Modified from Nace, 1971 and various sources
(Reproduced with permission from Chapman, D., 1992)

nitric and sulphuric acids as a result of industrial pollution. When rain falls on the land further changes in quality take place as a result of the twin processes of *erosion* and *solution*. Thus the composition of water in rivers and lakes, in terms of both dissolved and suspended matter, closely reflects the geology of the river basin. Such differences can have major influences on the aquatic flora and fauna, as exemplified by the contrast between highly calcareous streams in the chalk downs of southern England and the 'soft' waters of many Welsh and Scottish streams. The former are highly productive and support richer, more diverse aquatic communities which are biologically very different from those in Wales and Scotland yet both sets of streams would be considered excellent for recreation, angling and water supply purposes.

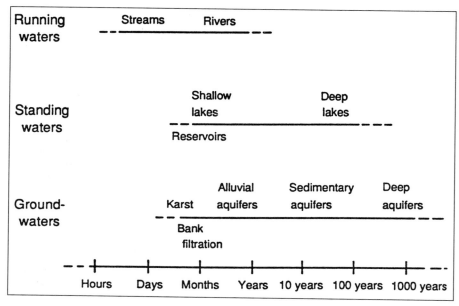

Fig. 1 Water residence times in inland water bodies
(After Meybeck et al. 1989; reproduced with permission from Chapman, D. 1992)

DEFINITION OF WATER QUALITY

There is such a wide range of factors which can be used to define water quality that no simple definition exists. Water which is of excellent quality for certain purposes may be totally unsuitable for other purposes. Thus drinking water would almost certainly be useless for boiler feed water because of its dissolved constituents, particularly calcium, magnesium and silicon, and would also be lethal to fish life because of its chlorine content.

Water *resources* are usually described in purely hydrological terms such as flow rate and current velocity in rivers or the depth, volume and retention period of lakes or reservoirs, while the water *quality* may be described for some purposes solely by physical and chemical parameters such as temperature, dissolved oxygen and hardness and, for other purposes, by biological measurements such as counts of microscopic plant and animal life.

However, the overall quality of the aquatic environment in lakes, rivers and coastal waters can only be *fully* described in terms of hydrology, physico-chemistry and biology and also by taking into account the location and timing of sample collection. Thus a sample taken at mid-stream and mid-depth will normally accurately reflect the quality across the whole river, but in a large deep lake samples taken at different depths and different points may show big differences in quality because of thermal stratification and changes in biological productivity between the inlet and outlet. Changes in oxygen, pH, carbon dioxide and other variables are often substantial during a 24-hour period due to the influence of solar radiation on temperature and photosynthesis. In estuaries and coastal waters tidal variations will also have an impact on water quality, and sampling programmes must take account of all these factors if an accurate description of water quality is to be obtained. In both inland and coastal waters there are, of course, seasonal differences in water quality due to the impact on biological processes of differing lengths of daylight, temperature and other factors.

PROPERTIES OF WATER

Density

It is well known that pure water is a liquid 775 times more dense than air at normal temperature and pressure, which freezes at 0° and boils at 100° Celsius. It is less well known that water is unique in the way that its density changes with temperature. Unlike all other substances, whose density increases continuously with falling temperature, water becomes heaviest at 4°C (strictly 3.94°) and then decreases slowly in density to 0°C. When freezing takes place the density decreases abruptly as a result of expansion so that ice is about one twelfth lighter than water at 0°C. This remarkable temperature/density relationship (illustrated in Fig. 2) is caused by the arrangement of water molecules which change from a tetrahedral pattern in ice to a closely packed spherical arrangement (like peas in a box) at the temperature of maximum density. Above this temperature normal thermal expansion accounts for the decrease in density. By reference to the curve in Fig. 2 a further important point should be noted, i.e. that the density changes increasingly rapidly with rising temperature, so much so in fact, that the decrease in density between 24° and 25° is thirty times greater than between 4° and 5°C.

This relationship between temperature and density is of critical importance to life in the aquatic environment. Firstly, the fact that water bodies such as lakes, ponds and the sea freeze from the surface downwards enables aquatic organisms to survive long winter periods of sub-zero temperatures which would be quite impossible if freezing took place from the bottom upwards, with the result that such water bodies would freeze solid. Aquatic organisms are thus exposed to much smaller temperature

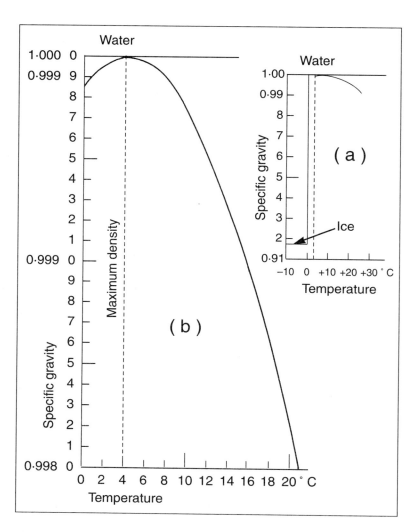

Fig. 2 (a) Dependence of specific gravity of water and ice on temperature. (b) Section of curve between 0° and 20°C with specific gravity scale magnified 20 times and temperature scale 5 times
(Reproduced with permission from Ruttner, F. 1953)

variations than their terrestrial counterparts and do not have to survive freezing conditions which limit the spread of many species.

Secondly, the biological and chemical processes in the deeper lakes are also profoundly influenced by density gradients which are induced by the warming effect of solar radiation on the surface layers. Wind-induced currents will tend to cause vertical mixing in shallow waters, but in deeper bodies wind energy is insufficient to bring about mixing of the light surface layers with the more dense lower layers, and such lakes become *stratified* into two district layers – a surface layer (the *epilimnion*) which is warm, well illuminated, highly productive and rich in dissolved oxygen as a result of photosynthesis, and a cold bottom layer (the *hypolimnion*) which is isolated from sunlight and where, consequently, only the processes of respiration and decomposition take place. The *hypolimnion* then becomes deficient in oxygen which is replaced by carbon dioxide. Anaerobic decomposition in the underlying sediments may also release substances such as methane, ammonia and sulphuretted hydrogen. In this way thermal stratification gives rise to chemical and biological stratification with totally different biochemical processes taking place in the two layers.

Between the *epilimnion* and the *hypolimnion* is a transitional layer known as the *thermocline* in which there is a sharp drop in temperature from the bottom of the epilimnion to the top of the hypolimnion (see Fig. 3). The depth of the thermocline in lakes in temperate regions can vary from 5 to 15 metres according to such factors as the seasonal temperature range, wind strength and pattern, and the depth and shape of the lake. In shallow lakes it is usually a temporary feature in summer since wind action can easily cause mixing of the water from top to bottom, but in deep lakes the density differences may be so great that stratification may last throughout the summer or even become permanent.

Thermal stratification is also a feature of coastal waters around the UK and Europe with a thermocline at a depth of about 10–12 metres. In shallow coastal waters (20–30 m) stratification may be short-lived because of wind-induced circulation, but in deep waters such as in the sea lochs of Scotland (100–200 m) stratification may last many months. Since little or no exchange of water takes place through the thermocline the bottom layers gradually lose oxygen and, after prolonged periods, may become anoxic with rising levels of ammonia, hydrogen sulphide and methane.

Stratification is also well known in estuaries as a result of the strong salinity differences between fresh and salt water, the two layers being separated by a discontinuity layer known as the *pycnocline*, where there is a sharp salinity gradient.

Thermal stratification can pose problems in water supply reservoirs because of adverse effects on water quality in the bottom layer. However, the problem can be

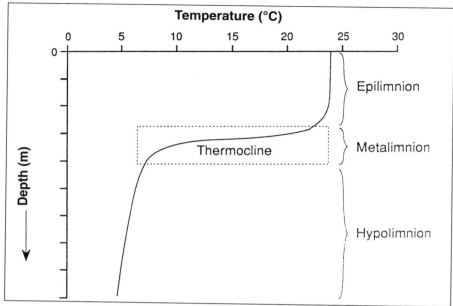

Fig. 3 Typical temperature profile of a stratified lake or reservoir showing division into three layers caused by heating surface water due to solar radiation in summer
(Reproduced with permission from Chapman, D. 1992)

overcome by angled jets of water (e.g. where the incoming water is pumped into the reservoir from a river) or by bubbles of compressed air released from pipes in the reservoir bottom to induce vertical mixing.

Viscosity

The viscosity of a liquid is a physical property which describes its resistance to flow, or the frictional resistance of the fluid to a body moving through the liquid. The viscosity decreases with rising temperature so that, at 25°C, microscopic bodies such as algae sink twice as fast as at 0°C. It also follows, from this relationship, that at summer temperatures the effect of wind velocity on surface currents and water mixing will be greater than during cold winter weather. The effect of varying salt content within freshwaters is not significant, but seawater is distinctly more viscous than freshwater.

Surface Tension

At the air-water interface water molecules are attracted to each other in an unbalanced way because there are no water molecules above them. This *surface tension* produces what is commonly known as the surface film because, for example, it can support the weight of small organisms such as pond skaters which are heavier than water. Surface tension affects light penetration (because the surface film reflects incident light at certain angles) and also influences the interchange of gases such as oxygen between water and the atmosphere. It is worth noting here that a specialized community of microscopic organisms, the *neuston*, mainly comprising bacteria, fungi and algae, is often found attached to the surface film.

Solvent ability

One of the most important properties of water in relation to water quality is its ability to dissolve almost all known substances with the consequence that pure water does not exist in nature – indeed it is very difficult to produce completely pure water in the laboratory. It is claimed, for example, that seawater contains (in solution) almost every known element, though many of these are found at extremely low concentrations.

Rainwater and surface waters in contact with the atmosphere generally have (in solution) the atmospheric gases oxygen and nitrogen in equilibrium with their concentration in the air. However, because of differing absorption coefficients, one litre of water at 20°C contains 6.4 millilitres (ml) of oxygen and 12.3 ml of nitrogen whereas in the atmosphere one litre of air contains 210 ml of oxygen and 790 ml of nitrogen. These concentrations and ratios vary significantly with temperature so that, for example, at 5°C one litre of water will contain 8.9 ml and 16.8 ml of these gases respectively.

As rainfall flows over and through the surface rocks in a river basin it takes into solution small amounts of the minerals with which it comes into contact, the resultant concentrations being a product of the relative abundance of the mineral and its solubility in water. At each point along a watercourse, an analysis of the water will reflect the nature of the river basin upstream and, of course, any man-made inputs such as from afforestation, agriculture, and piped discharges from towns and factories. Pumped discharges from mine workings and overflows from abandoned mines may also have a considerable impact due to underground processes which release acids and minerals, such as iron salts, into solution.

Erosion

Another important property of water is its power to erode soils and rocks and to transport the products of erosion downstream, often for vast distances. Many forces are involved in this process, i.e. the ability of water to dissolve rocks together with the fracturing of rocks by repeated freezing and thawing in mountain areas. Water, as ice in the form of glaciers, carves out huge valleys and the rivers act as huge transport systems, taking material which starts as boulders, and is progressively converted into smaller rocks, pebbles, gravels and, eventually, sands and clays which are deposited to produce alluvial plains. Much of this material is carried in fine suspension into lakes and estuaries where, because of reduced velocities, it settles out to form mudbanks and deltas. Most of the suspended and dissolved material is thus the result of erosion by water, though a proportion of particulate matter is derived from windblown sources, especially in desert areas.

Optical Quality

The ability of water to transmit and absorb solar radiation is another important aspect of water quality. Light is of profound importance as a driving force for most of the physico-chemical and biological processes taking place in all water bodies, therefore any detailed studies of environmental water quality should include its *optical quality* – affecting both *clarity* and *colour* which constitute the principal components of its 'visual' quality or 'appearance'.

The visual appearance of water bodies, and of samples of water taken from them, can tell us quite a lot about their quality before any analyses are carried out. The general public, particularly anglers and water sports enthusiasts, will often give early warnings of pollution incidents by noticing discoloration, lack of normal transparency or the presence of surface films. Experts in water quality can make useful deductions about the quality of lakes and rivers by visual inspection, which may indicate the degree of productivity and even the types of planktonic algae which are present.

Colour and turbidity together determine the transparency of water, i.e. the depth to which surface light is transmitted. However, of the vast array of dissolved, colloidal and particulate materials which may be present, relatively few absorb or scatter light. Thus it is of interest to note that the high concentration of dissolved salts in seawater has almost no optical effect, and many seawaters are virtually indistinguishable optically from pure freshwaters.

Contrary to popular belief, pure water is not colourless but has a very pale blue colour. However, most natural waters are affected by the presence of algae, zooplankton, dissolved organic, mineral matter and suspended particulate matter, and can range from greenish blue, through green to greenish yellow, yellow or brown. Here we should distinguish between the true colour of water and the apparent colour, the former being due to dissolved matter and the latter to colour resulting from suspended matter.

The optical quality of water has been a neglected area in water pollution control despite being long recognised as of fundamental importance in academic studies in oceanography and limnology. In recent years, criteria for measurement of optical water quality have been developed and may become increasingly useful in the water quality management of water supply reservoirs.

Turbidity

Turbidity results from the scattering of light by particulate (i.e. undissolved) matter in suspension. Rivers are often turbid, due to the presence of soil particles washed in from the river valley after rainfall, while the turbidity of estuarine waters is more often due to the resuspension of bottom muds as a result of tidal scouring. In coastal waters turbidity is more often due to the effect of wave action on the shore line.

Lake waters are usually free of turbidity since suspended materials usually sink to the bottom. However, eutrophic lakes (i.e. those which are rich in nutrients such as phosphate and nitrate) may be turbid and coloured in shades of green or blue/green due to the presence of large numbers of microscopic algae or bacteria.

Turbidity is a key factor influencing the productivity of natural waters because, by reducing light penetration, it can restrict photosynthesis, i.e. primary production on which all life depends.

CLASSIFICATION OF WATER BODIES

The biological and chemical quality of water is closely associated with the type of water body from which it is derived. For example, lake waters generally have a low content of suspended mineral matter but a rich growth of microscopic plant and animal life while, conversely, rivers have a higher content of silt in suspension and lower concentrations of living organisms. It is useful, therefore, to classify water bodies according to their hydrodynamic properties as follows:

Rivers

Rivers are characterized by unidirectional currents with high velocities in the range 0.1 to 1.0 metres/second. The discharge is highly variable in time, depending on climate and basin topography. River waters are generally well mixed vertically due to turbulence, but lateral mixing of tributaries or effluent discharges may only be achieved over long distances. In large rivers such as the Nile or Amazon, full mixing of tributaries may not occur until well over 100 to 200 kilometres downstream from a confluence.

Lakes

Lakes have low current velocities, which may be multidirectional, in the range of 0.001 to 0.01 metres/second. Unlike rivers, where water residence times are short, the residence time can range from several weeks to hundreds of years. As seen earlier, lakes may be seasonally or permanently stratified with important consequences for water quality.

Groundwaters

Groundwaters are characterized by very slow and stable flows in the range of 10^{-10} to 10^{-3} metres/second related to the permeability and porosity of the aquifer. Mixing is poor and long residence times are usual, although greatly reduced in aquifers used for water supply. It has been calculated that the volume of water in aquifers in England and Wales alone is about 16 billion cubic metres, most of which is in deep aquifers with little inflow or outflow. By contrast, the accessible groundwater for the whole of the UK is between 1.5 and 2.5 billion cubic metres for which the average residence time is about two months. However, in fine-grained aquifers such as chalk, groundwater may remain for years or even centuries before it is discharged naturally or abstracted. Long residence times mean that groundwater takes far longer than surface water to recover from pollution and may need special protection measures.

Intermediate water bodies

Flood plains: these are intermediate between rivers and lakes and are seasonal.
Reservoirs: these are also intermediate between rivers and lakes depending on the pattern of operation (filling and emptying).
Marshes: intermediate between lakes and aquifers.
Alluvial and karstic aquifers: intermediate between rivers and groundwaters with slow (alluvial) and rapid (karstic) flow.
The relative residence times in these water bodies are shown in Table I.

It is important to emphasize that effective water quality monitoring systems can only be devised and interpreted by fully understanding the hydrological characteristics of the water body under consideration.

The same principle also applies to estuaries and coastal waters. Estuaries, in particular, are highly complex hydrodynamically because of the mixing of inflowing freshwater with seawater and the impact of tidal movement. No two estuaries are alike and the pattern of water mixing will depend on many factors including size, shape, depth (especially any artificial deepening) and the tidal range. Thus estuaries such as the Clyde and Thames, which have been greatly deepened by dredging for shipping, tend to function as large settling basins in which the water moves very slowly back and forth with the tide so that the inflowing river water may take weeks to reach the open sea. As a result, at low river flow in summer, the organic wastes may so deprive the water of oxygen that the estuary becomes impassable to migratory fish – a situation which has only been overcome in both estuaries after more than one hundred years.

2. THE MEASUREMENT OF WATER QUALITY

PHYSICAL CHARACTERISTICS

Temperature

Temperature is a key characteristic of all water bodies because it influences physical, chemical and biological processes. Most water bodies exhibit both seasonal and diurnal variations in temperature, which are related to climate and geographical location. Rivers and shallow lakes may have a uniform temperature from top to bottom, but deep lakes and coastal waters may become stratified into warm surface layers above cold lower layers as temperatures rise in the spring and summer months. In many waters the rapid development of algal blooms is accompanied by high rates of photosynthesis which result in high concentrations of dissolved oxygen and considerable fluctuations in dissolved gases, while nutrients such as phosphate and nitrogen are consumed by the algae and higher plants. In our temperate climate the annual range of water temperature in surface waters is usually within the range from 0°C to 25°C and rarely reaching 30°C.

Temperature is always measured *in situ* at the time of sampling by means of a thermometer or thermistor. In deep waters a thermistor or reversing thermometer attached to a water bottle is essential though, if the depth is not excessive, measuring the temperature of the sample as soon as it is brought to the surface may be adequate. (A reversing thermometer 'fixes' the temperature at depth by means of a weight sent down the cable which inverts the thermometer thus breaking the mercury column.)

Transparency

Transparency is a useful guide to biological productivity in lakes and reservoirs and is measured in the field, usually from a boat, by lowering a white circular disc 200 to

300 mm in diameter (known as a Secchi disc) on a graduated cable and observing the depth at which it disappears from sight. It is then slowly raised and the depth at which it reappears is noted. The mean of these two depths is recorded as the depth of transparency.

The transparency of natural waters can vary substantially, but in the UK reservoirs and lakes the range is usually from 0.5 m to 15 m. Very high values of over 50 m have been observed in tropical seas and in some crater lakes. In European alpine lakes, values as high as 20–25 m have been observed.

Colour

The true colour of a sample of water can only be measured in the laboratory after filtration to remove any turbidity due to particles in suspension. 'Apparent colour' can be recorded before filtration and may be sufficient for some purposes. Samples which contain iron may become turbid and coloured orange after collection due to precipitation of ferric hydroxide (e.g. acid minewaters) and it is therefore important to note the appearance at the time of collection as well as at the time of analysis.

The true colour is measured after filtration, either visually or by a photoelectric method, by comparison with standard solutions of potassium chlorplatinate and crystalline cobaltous chloride.

Turbidity

Turbidity in water arises from the scattering of light by suspended solids and depends not only on the concentration but also on the size, shape and refractive index of the particles. Thus, although turbidity measurements are used as a guide to the concentration of materials in suspension, the results must be treated with caution since samples with the same turbidity could contain differing concentrations depending on the size, shape and refractive index of the suspended particles.

Turbidity measurements are of particular interest to managers of water treatment plants as a quick and simple guide to water quality. Continuous measurement of turbidity is used in process control, for example in monitoring the performance of sedimentation tanks.

It is important that turbidity measurements are made on the same day that samples are collected although, if necessary, samples can be kept for not more than 24 hours in the dark. The usual method for determining turbidity is by comparison with standard formazin suspensions using a nephelometer, though an absorptiometer can be used. A nephelometer determines the ratio of the intensity of scattered light at 45° to the main path, to that of the transmitted light, while an absorptiometer measures the ratio of the intensity of transmitted light to that of incident light.

It should be noted that different nephelometers, particularly if they differ widely in design, can give differing results for the same samples even when calibrated with the same formazin standards. For these and other reasons, turbidity measurement by nephelometer is not recommended for accurate, academic studies of the optical properties of natural waters.

Suspended Solids

Almost all surface waters contain suspended particulate matter which is derived from natural weathering in the catchment, erosion of the river bed as a result of current velocity, or by wave action on the shores of rivers, lakes and coastal waters and also from windblown sources. A wide range of land-use activities such as agriculture, forestry, and road building can cause the discharge from diffuse or non-point sources of large amounts of suspended solids. Many industries, including paper mills, mines, quarries, and manufacturing industries make direct, piped discharges of polluting materials including suspended solids.

Suspended solids, by settling out on the river lake or sea bed, can be harmful to bottom living organisms and prevent the breeding of fish (such as salmon and trout) by stopping the free flow of oxygenated water through the gravel in which the fish lay their eggs. Suspended solids also reduce light penetration and thus inhibit the growth of algae and larger plants.

By common agreement it is widely accepted that *suspended solids* (SS) or *particulate matter* (PM), often described as *total suspended solids* (TSS), comprise particles which are retained by a filter of 0.45 µm pore diameter. (Accordingly *dissolved matter* includes very fine particles less than 0.45 µm in diameter, such as colloidal matter.) Total suspended solids are determined in the laboratory by passing a suitable volume of water through a filter of 0.45 µm pore size and weighing the retained solids after drying to constant weight at 105°C.

TSS is one of the most variable of water quality characteristics, especially in rivers. For example, in the River Nile the TSS can vary from less than 10 parts per million in the dry season to over 10 000 ppm in the flood season. In Europe the annual variation can range from 1 ppm to over 1000 ppm. For this reason, annual averages need to be treated with great caution unless very frequent sampling is carried out.

Total Dissolved Solids

Total dissolved solids (TDS) are determined by passing a measured volume of the water sample through a filter of 0.45 µm pore size and evaporating the filtrate to constant weight in an evaporating dish at 105°C. This determination needs to be carried out as soon as possible after sample collection.

Conductivity

Electrical conductivity provides a simple and speedy way of estimating the concentration of dissolved inorganic salts. The method is particularly useful in relatively clean rivers, lakes and groundwaters which are potential sources of drinking water. However, organic compounds, because they do not dissociate into ions or dissociate only very slightly, are poor conductors, and conductivity does not provide a good measure of their concentration. The test is best carried out in the field using a conductivity meter which automatically compensates for variations in temperature. (Conductivity measurements are ideally carried out at the standard temperature of 20°C, but if automatic compensation is not available a correction can be made provided that the temperature is noted.)

Conductivity is measured in siemens per metre or, in pure lakes and rivers, in millisiemens per metre. Fresh distilled water has a conductivity of 0.1–0.2 millisiemens/metre but will increase on exposure to the atmosphere owing to absorption of carbon dioxide to about double this value. The values found in most natural waters are in the range of 5–50 mS/m, although in well-mineralized waters conductivity may rise to 100 or higher and in very polluted waters to 1000 mS/m.

pH

The pH scale is a convenient way of measuring the acidity or alkalinity of a solution and extends from 0 (extremely acid) to 14 (extremely alkaline), while the mid-point pH 7 corresponds to exact neutrality at 25°C. The pH is defined as the logarithm to base 10 of the reciprocal of the hydrogen ion concentration. Because this is a logarithmic scale, 1 unit represents one order of magnitude so that water at pH 4.0 is 10 times more acidic than at pH 5.0 and 100 times more acidic than at pH 6.0. Likewise pH 8.0 is 10 times more alkaline than pH 7.0.

In natural waters the pH is controlled by the relationship or equilibrium between carbon dioxide, bicarbonate and carbonate ions (Fig. 4) and can be strongly influenced by biological activity. Thus photosynthesis by algae and other plants during daylight hours utilizes carbon dioxide and the pH may rise sharply, while by night the main biological activity is respiration which produces carbon dioxide, and the pH decreases accordingly.

The natural pH of rain is around 5.6, but industrial emissions of sulphur dioxide together with oxides of nitrogen from industry and cars lead to the formation of sulphuric and nitric acids, with the result that rainfall has become much more acidic

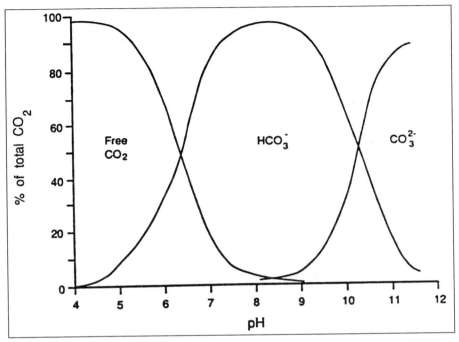

Fig. 4 Relative proportions of free carbon dioxide (CO_2), bicarbonate (HCO_3^-) and carbonate (CO_3^{2-}) in water in relation to pH
(Reproduced with permission from Chapman, D. 1992)

during this century, average values ranging from 4.3 in the south-east of England to 4.7 in the north-west of Scotland. The dry deposition of sulphur dioxide during dry spells worsens the situation by making the runoff even more acidic with the resultant acidification of lakes and rivers in 'sensitive' areas. These are areas where there is little lime (calcium carbonate) in the rocks and surface soils to neutralize the acidity in the runoff. Acidification has damaged many waters in Wales and Scotland resulting, in some cases, in the total loss of fish life. Recent reductions in emissions have begun to rectify the situation, but full recovery will require tighter controls and is likely to take many years. Some waters are naturally acidic in mountain areas due to humic acids, but not as acidic as those affected by industrial acidification.

pH is now measured accurately by means of a glass electrode in an instrument which measures and compensates for errors due to temperature differential. A wide range of instruments is available, many of which are specially designed for field use. pH should always be measured *in situ* or at least as soon as the sample is taken because it can change fairly rapidly due to a range of natural factors.

CHEMICAL CHARACTERISTICS

As mentioned earlier, water has a remarkable solvent ability such that pure water is unknown in the natural environment and difficult to produce in the laboratory. Water reaching the earth's surface as rain, hail or snow already contains small amounts of dissolved gases (oxygen, nitrogen and carbon dioxide) absorbed in its passage through the atmosphere, but the subsequent dissolving of minerals from the surface rocks largely determines the streamwater chemistry. Rivers and lakes can, therefore, vary enormously in character depending on the physical and geological nature of the catchment. Streams in mountain areas of Wales and Scotland, where the rocks tend to be igneous and hard wearing, are generally acidic, low in dissolved minerals and regarded as 'soft' for washing purposes. By contrast, waters from catchments where sedimentary calcareous rocks predominate, such as the limestone Cotswold hills or the chalk downs of Sussex, are generally alkaline, rich in dissolved minerals and considered 'hard' for domestic purposes.

It has long been recognized that the communities of plants and animals which live in water are highly sensitive to their physico-chemical environment, and even small differences in the chemical components are reflected in the structure and species diversity of these communities. Even bigger differences are found when water is polluted by sewage and industrial wastes.

The chemical components of natural waters can be conveniently grouped in five categories: particulate inorganic substances; dissolved inorganic substances; particulate organic substances; dissolved organic substances; and dissolved gases.

Particulate Inorganic Substances

The main natural source of minerals, which are carried in suspension by rivers, is the mechanical erosion of rocks and soils brought about by wind, running water, frost, ice movement and land slips. Such processes are most active in high mountain areas and the composition of sediment in mountain streams closely corresponds to the geology of the catchment.

In the middle and lower reaches of most rivers the minerals in suspension are increasingly derived from soil particles washed in after rainfall on the floodplains. The chemical composition of these soil particles is often markedly different from the parent rocks from which they were derived because soluble components, such as sodium, potassium, calcium and magnesium, have been washed out and transported down river in solution leaving the soil particles correspondingly richer in the insoluble ones such as aluminium, iron and manganese. The silt in suspension changes in chemical composition in exactly the same way as it is transported downstream, so that the sediments which are deposited each year in the flood plain and, finally, in the estuary, are considerably modified.

In lakes and marine waters (and, to a limited extent, in rivers) particulate matter is also derived from biochemical processes such as photosynthesis which cause the precipitation of calcium from solution as calcite (calcium carbonate).

Dissolved Inorganic Substances

As soon as water falls on the earth's surface it starts to dissolve minerals according to the acidity/alkalinity of the water, the relative solubilities of the minerals and other factors (including the temperature and concentrations of ions already in solution).

Rainwater contains many ions in solution (Table 2) and of these it has been shown that sodium, potassium, calcium, magnesium and chloride are derived mainly from particles in the atmosphere while sulphate, ammonium and nitrate ions are derived mainly from atmospheric gases. In coastal areas of continents, marine salts (mostly

Table 2. Typical concentrations of major ions in rainfall (mg/l)

Ion	Continental rain	Marine and coastal rain
Na^+	0.2–1	1–5
Mg^{2+}	0.05–0.5	0.4–1.5
K^+	0.01–0.5	0.2–0.6
Ca^{2+}	0.2–4	0.2–1.5
NH_4^+	0.1–0.5	0.01–0.05
H^+	pH 4–6	pH 5–6
Cl^-	0.2–2	1–10
SO_4^{2-}	1–3	1–3
NO_3^-	0.4–1.3	0.1–0.5

Source: Berner and Berner, 1987.
(Reproduced with permission from Allen, J.D., 1995)

sodium chloride) predominate, while further inland calcium sulphate or bicarbonate become increasingly dominant.

Studies have shown that, as a global average, 54% of rainwater falling on the earth's surface is evaporated, leaving 46% as runoff so that the concentration of ions in river water is over double that derived from rainwater. It has also been shown that surface waters contain, on average, twenty times the concentration found in rain, which confirms what one would expect, i.e. that most of the dissolved inorganic material in rivers and lakes originates from the weathering of rocks and soil and from man-made sources. In North America it has been demonstrated that 10 to 15% of calcium, sodium and chloride comes from rainfall in contrast to about one quarter of the potassium and half the sulphate. It is interesting to note that virtually none of the silicate or bicarbonate in river waters has been found to come from rain. With regard to man-made inputs, it is known that many pollutants emitted to the atmosphere from factories or evaporated from the land are washed out by rainfall and finish up in the rivers. A notable example of this is the occurrence of acid episodes in some rivers following heavy rainfall, which have resulted in the death of fish eggs and young fish.

The dominant dissolved substances in river waters are shown in Table 3. This table is of particular interest because it makes an estimate of the natural values (to exclude pollution) and, by comparing these with the actual values, it is possible to make a crude estimate of the inputs resulting from human activity. In Europe, 34% of the total dissolved solids originate from man-made inputs compared with 6.4 % for North America and 4.5% for Africa. The contrast is even more striking in the case of chloride (a good indicator of the proportion of treated sewage effluent in rivers) where, in Europe, 76% of the input is from human sources compared with 24% for North America and 17% for Africa. To a considerable extent these values are a reflection of population density in the respective continents.

The inorganic constituents are conveniently grouped as shown in Table 4.

Major Ions

Sodium

As one of the most abundant elements in the earth's crust, sodium is present in all natural waters and particularly abundant in seawater. The natural concentration in surface waters is usually below 10 mg/l, and rarely above 50 mg/l. Sodium is of concern in irrigation waters in arid climates where salt levels may increase due to evaporation. Sodium may then replace the calcium and magnesium in the soil and thus degrade the soil structure. Analysis is normally carried out by atomic emission spectrophotometry.

Table 3. Chemical composition of river water (mg/l)[a]

	Total dissolved solids	Ca^{2+}	Mg^{2+}	Na^+	K^+	Cl^-	SO_4^{2-}	HCO_3^-	SiO_2
World average									
Actual	110.1	14.7	3.7	7.4	1.4	8.3	11.5	53.0	10.4
Natural	99.6	13.4	3.4	5.2	1.3	5.8	8.3	52.0	10.4
North America									
Actual	142.6	21.2	4.9	8.4	1.5	9.2	18.0	72.3	7.2
Natural	133.5	20.1	4.9	6.5	1.5	7.0	14.9	71.4	7.2
South America									
Actual	54.6	6.3	1.4	3.3	1.0	4.1	3.8	24.4	10.3
Natural	54.3	6.3	1.4	3.3	1.0	4.1	3.5	24.4	10.3
Europe									
Actual	212.8	31.7	6.7	16.5	1.8	20.0	35.5	86.0	6.8
Natural	140.3	24.2	5.2	3.2	1.1	4.7	15.1	80.1	6.8
Africa									
Actual	60.5	5.7	2.2	4.4	1.4	4.1	4.2	26.9	12.0
Natural	57.8	5.3	2.2	3.8	1.4	3.4	3.2	26.7	12.0
Asia									
Actual	134.6	17.8	4.6	8.7	1.7	10.0	13.3	67.1	11.0
Natural	123.5	16.6	4.3	6.6	1.6	7.6	9.7	66.2	11.0
Oceania									
Actual	125.3	15.2	3.8	7.6	1.1	6.8	7.7	65.6	16.3
Natural	120.6	15.0	3.8	7.0	1.1	5.9	6.5	65.1	16.3

[a] Actual concentrations include inputs from human activity. Natural values are corrected to exclude pollution.
Source: Berner and Berner, 1987.
(Reproduced with permission from Allen, J.D., 1995)

Table 4. Dissolved inorganic constituents of natural waters

Major ions	Minor ions	Trace elements	Gases
Sodium Potassium Calcium Magnesium Bicarbonate Sulphate Chloride	Nitrate Ammonium Phosphate Silicate	Iron; copper; cobalt; boron; manganese; molybdenum; zinc; aluminium	Oxygen; nitrogen; carbon dioxide

Source: Modified from Whitton, B.A., 1975.

Potassium

Rocks containing potassium are fairly weather-resistant, and concentrations are generally low in rivers and lakes except in agricultural areas where fertilizers may give rise to high levels. Industrial discharges can also cause problems. Analysis is best carried out by atomic emission spectrophotometry.

Calcium

Calcium is widely distributed in surface rocks, usually in the form of carbonate or sulphate, e.g. limestone, chalk and gypsum. Natural waters usually contain less than 15 mg/l, but where calcareous rocks abound may reach 20–30 mg/l. In the presence of carbon dioxide, calcium compounds are stable in solution but high temperatures and the uptake by plants of carbon dioxide during photosynthesis may cause the precipitation of calcium as carbonate. Many lakes have extensive marl beds due to this process which can also occur in rivers. Calcium is vital for all living organisms and, in particular, for the external shell or exoskeleton of aquatic invertebrates. Concentrations below 3–5 mg/l can greatly reduce the populations of crayfish, insects and aquatic snails.

Calcium and magnesium, which are almost invariably present together, are responsible for 'hardness' in water, which can be defined as the ability of the water to precipitate calcium and magnesium salts of fatty acids from soap solutions. Heating precipitates calcium and magnesium carbonate which are responsible for boiler scale and the furring up of domestic hot-water systems. However, a thin coating of calcium carbonate is desirable to protect against corrosion, and a low dose of lime is sometimes added to soft waters during treatment for public supply, particularly in areas where the presence of old lead pipes may result in undesirable levels of lead in solution.

Two methods of analysis are available, i.e. by titration with EDTA (ethylenediamine tetra-acetic acid) or by atomic absorption spectrophotometry.

Magnesium
Like calcium, magnesium is an essential element for living organisms and universally present in natural waters. Also like calcium it is normally present in organic matter and in organometallic compounds. Concentrations in natural waters are usually in the range of 1 to 100 mg/l. As with calcium, magnesium can be determined by titration with EDTA or by atomic absorption spectrophotometry.

Carbonates and bicarbonates
All natural waters (rivers, lakes, estuaries and marine waters) contain (in solution) carbon dioxide, carbonic acid, carbonate and bicarbonate ions in a buffered system, which is vitally important to the health of the aquatic flora and fauna.

Carbon dioxide is derived from the atmosphere by absorption at the air-water interface, from the respiration of aquatic organisms and from the bacterial decomposition of organic matter, either in suspension or within the bottom sediments. The weathering of rocks contributes carbonate and bicarbonate salts and contributes about half the carbonate and bicarbonate concentration, whereas in areas of non-carbonate geology the carbonate and bicarbonate is entirely derived from the atmosphere and biogenic carbon dioxide.

In clean, unpolluted waters the acidity/alkalinity (or pH) is mainly controlled by the relationship between the carbon dioxide, bicarbonate and carbonate ions as illustrated in Fig. 4. Waters which are rich in calcium bicarbonate tend to be slightly alkaline (around pH 8.3) and well buffered, and therefore resistant to strong changes of pH. Such streams, typical of chalk or limestone catchments, are, for example, fairly resistant to the impact of acid rain. However, where they are enriched with nutrients and subject to heavy algal growth, high rates of photosynthesis in summer may utilize most of the available carbon dioxide with the result that the pH rises to between 9.5 and 10.5, which can kill fish.

It should be noted that photosynthesis plays a major role in the precipitation of calcium carbonate in lakes and marine waters. River waters are often supersaturated with carbon dioxide and, on entering lakes, this escapes to the atmosphere until equilibrium is regained – a process which is considerably speeded up by the photosynthesis of algae which consume carbon dioxide and release oxygen. It is essentially the same mechanism which is responsible for reducing the hardness of water in storage reservoirs; likewise, the loss of carbon dioxide to the atmosphere from dripping water is responsible for the formation of stalactites and stalagmites in limestone caves.

In very pure waters, deficient in calcium carbonate, as found in Wales, the Lake District and the Highlands of Scotland, the results of acid rain may seriously depress the pH to 4.5 or less – resulting in the death of fish eggs and fry. In this case the deaths are not directly caused by the acidity but result from the toxic metal aluminium (one of the most abundant elements in soil) which is leached from the soil under acid conditions.

Marine waters are very well buffered and thus the pH range is small.

Concentrations of carbonate, bicarbonate and carbon dioxide are usually calculated from measurements of the pH (by glass electrode), alkalinity (by titration), temperature and ionic strength.

Sulphate

Sulphate is widely distributed in the surface rocks and is highly soluble and therefore easily leached from gypsum (calcium sulphate), sodium sulphate and some shales. Mine drainage may contain sulphates due to the oxidation of pyrite. Sea water contains sulphate, and rainfall in coastal areas can contribute significantly to sulphate levels in rivers and lakes. In natural waters, concentrations range from 1 to 80 mg/l, but pollution from industrial sources may result in much higher levels. Natural hot spring waters may contain several thousand mg/l.

Analysis can be carried out by the gravimetric determination of barium sulphate after ignition or by EDTA titration.

Chloride

Chloride is universally present in freshwaters which are derived from the weathering of sedimentary rocks and rainfall. As seen in Table 3, the average natural concentration in rivers is well under 10 mg/l. However, high levels can occur naturally due to weathering of rock salt and, more often, due to sewage and industrial inputs. Sea water contains around 20 000 mg/l and, in coastal areas, high concentrations can result from the intrusion of sea water into the groundwater.

Chloride is both chemically and biologically unreactive and is sometimes used as a tracer in nutrient-release studies. Because it is unreactive, samples for analysis need no special treatment or preservative and can be stored at room temperature. Analysis can be carried out by titration with silver nitrate using either a colorimetric or potentiometric end point. Direct determination can also be undertaken with chloride-sensitive electrodes.

Minor Ions

Nitrate, phosphate and silicate, though described here as minor ions in terms of concentration, are of vital importance in aquatic ecosystems as nutrients for plant growth. They are of particular importance to the growth of phytoplankton (the primary producers) and may limit productivity at low concentrations while high levels may cause undesirably high growths of macrophytic plants and algae.

Nitrogen and its compounds

Nitrogen is one of the essential 'building blocks' for all living organisms as a major constituent of protein and genetic material such as DNA. Bacteria and plants utilize inorganic nitrogen from the surrounding water to produce organic matter such as protein within their cells, while animals such as zooplankton and fish derive their protein from plants and may build this into more complex protein and organic matter. When these organisms die, a range of micro-organisms (by a process of ammonification) convert organic compounds such as the various proteins to ammonia. Under aerobic conditions, nitrifying bacteria such as *Nitrosomonas* and *Nitrobacter* convert ammonia through nitrite to nitrate, but under anoxic conditions denitrifying bacteria carry out the reverse process, converting nitrate and nitrite to ammonia and eventually to gaseous nitrous oxide and elemental nitrogen. These transformations form part of the same well-known nitrogen cycle found in terrestrial eco-systems.

In natural, unpolluted rivers and lakes, nitrogen is present as nitrate, generally at concentrations below 0.1 mg/l as nitrate nitrogen. As a result of human activity, many rivers in the UK contain up to 5 mg/l, while in areas of high agricultural production such as East Anglia rivers approach or exceed the WHO maximum recommended drinking water value of 11.3 mg/l.

In natural waters, levels of ammonia and of nitrite are usually below 0.1 mg/l because nitrifying bacteria ensure that any residues are oxidized to nitrate.

Phosphorus compounds

Like nitrogen, phosphorus is an essential element for plant growth, being a constituent of protein, phospholipids and nucleic acids. It is present in the aquatic environment in the form of orthophosphate, polyphosphate and organically bound phosphates. Phosphorus is known to play an important role in respiration and photosynthesis. Because it is rapidly taken up by algae, concentrations rarely rise significantly and large algal blooms may depress phosphorus to almost vanishing point.

In natural waters, phosphorus is usually found in the range of 0.005 to 0.02 mg/l as phosphorus, but concentrations one to two magnitudes higher may occur in polluted waters.

In water quality assessment it is desirable to measure the different forms of phosphorus, usually total phosphorus, orthophosphate phosphorus and total inorganic phosphate. Dissolved forms of phosphorus can be measured after filtration and particulate phosphorus estimated by difference. All results are reported in terms of phosphate phosphorus.

Silicate

Silicon, after oxygen, is the most abundant element in the earth's crust and a major constituent of igneous and metamorphic rocks, clays, feldspars and quartz. However, crystalline silica (quartz), present in many igneous rocks and sedimentary sandstones, is largely insoluble and it is believed that dissolved silica in surface waters derives from the chemical weathering of aluminium silicates by complex processes.

Various forms of silicon may be present in surface waters but it is usually measured and reported as the oxide, silica. In most natural waters the concentration of silica lies between 1 and 30 mg/l, but brackish waters may contain much higher levels (up to 1000 mg/l).

Silicate is an essential nutrient for one specialized group of algae, the diatoms, and because this is a major group whose population growth is closely dependent on silicate, measurement of this parameter is important for management purposes.

Analysis is usually carried out by a colorimetric method. It is essential to use plastic containers to prevent leaching of silicate from glass.

Trace Inorganics

The eight metals listed in Table 4 are those which are almost universally found in natural waters and which are known to be of importance as micronutrients. Most are present at concentrations of between one nanogram and one microgram per litre (i.e. between one part in a million million parts and one part in a thousand million parts of water by weight). However, as mentioned later, as a result of pollution, concentrations of these metals in surface waters may become dangerous for aquatic organisms and for human health.

Chemical analysis at these low levels requires specialized methodology involving highly trained staff. Atomic absorption spectrometry with electrothermal atomization is accurate but expensive and slow, and is now mainly replaced by atomic emission spectroscopy or, preferably, inductively coupled plasma atomic emission spectroscopy, which has a high capital cost but allows the rapid determination of elements (Plate 1).

Particulate Organic Substances

Particulate organic matter in aquatic systems includes living microscopic organisms such as phytoplankton, zooplankton and bacteria, dead organisms, breakdown material, e.g. from decaying leaves and larger animals, and faecal material. In smaller streams, especially in forested areas, landward sources provide much of this input in the form of falling leaves and twigs and windblown pollen and seeds. Together with the particulate inorganic matter it comprises the suspended solids referred to earlier; also, with dissolved organic matter it provides an important energy source, especially in rivers.

Particulate organic matter is measured as particulate organic carbon (POC) by a variety of methods, all of which are based on the oxidation of the carbon in the sample (after filtration and drying) to carbon dioxide. The oxidation is performed by combustion in an electric furnace or by chemical reaction or by ultraviolet radiation. The carbon dioxide produced is then measured using a specific electrode or by thermal conductance or a volumetric method.

Dissolved Organic Substances

These are derived from the breakdown of solid organic matter by natural biological processes, some of which take place in the aquatic environment and others which take place in soil and groundwater. Dissolved organic matter may comprise a wide range of organic compounds such as humic acids, fulvic acids and hydrophilic acids, hydrocarbons, fatty acids, fats, waxes and oils. In polluted waters an enormous range of substances is potentially involved, such as pesticides (organochlorine and organophosphorus compounds), phenols and polyaromatic hydrocarbons. As with particulate organic matter, these substances are measured as dissolved organic carbon (DOC).

In most rivers and lakes DOC comprises the greater part of the total organic carbon (TOC) and is in the range of 1 to 20 mg/l in unpolluted waters. Only in turbid rivers or during flood conditions, when much organic material is swept into rivers and lakes, does POC exceed DOC in concentration. DOC can be determined as for POC by analysing water which has passed through a filter of 0.7 µm pore size.

Total organic carbon (TOC) is determined without filtration. Determinations for the wide range of natural organic compounds are outside the scope of this volume.

Gases in Solution

As explained earlier, oxygen and nitrogen dissolve in water in different proportions from those in the atmosphere. However, because it is chemically inert, nitrogen is not important in water quality studies and is not normally determined. The important gases from a water quality standpoint are oxygen and carbon dioxide.

Table 5. Solubility of oxygen in pure water at sea level

Temperature (°C)	Dissolved oxygen (mg/l)	Temperature (°C)	Dissolved oxygen (mg/l)
0	14.63	15	10.07
5	12.77	20	9.08
10	11.28	25	8.26

Oxygen

Oxygen is vital to all living organisms other than those which thrive under anoxic conditions and, because only small amounts are present in water even under saturated conditions, the concentration of dissolved oxygen is a key measure of the health of aquatic ecosystems. Fish are particularly sensitive to dissolved-oxygen concentration, especially game fish such as trout and salmon.

The solubility of dissolved oxygen in pure water (in equilibrium with the atmosphere at sea level) is shown in Table 5.

At water temperatures experienced in temperate climates, it will be seen that pure water in contact with the atmosphere will have a maximum dissolved-oxygen content of between 14.63 mg/l at 0°C and 9.08 mg/l at 20°C. Water holds progressively less oxygen at higher altitudes, e.g. on the top of mountains, and also with increasing salt concentration. However, the amount of dissolved salts in normal healthy rivers has no significant effect, and altitude only needs to be taken into account for sites above 500 metres. Seawater, however, holds appreciably less oxygen – about 78 to 80 per cent of the freshwater values between 0 and 20°C. In estuaries and coastal waters, salinity is therefore an essential determinand when assessing water quality.

In lakes, coastal waters and to a lesser extent in rivers, dissolved oxygen can show marked diurnal and seasonal variation – especially in eutrophic (highly productive) water bodies. In strong sunlight in summer, the surface layers may become supersaturated, with oxygen concentrations of 150–200% saturation being not unknown, while, during the hours of darkness, respiration of the plankton may reduce the dissolved oxygen to saturation values of 50–80%. In extremely eutrophic waters, especially in conjunction with a degree of sewage pollution, the dissolved-oxygen concentrations may decrease to near zero, thus causing fish deaths.

When studying the diurnal changes in water bodies it is found that photosynthesis by day not only oxygenates the water but, by utilizing carbon dioxide, raises the pH of the water. Not only is there a daily pattern of water quality change associated with photosynthesis and respiration but also a seasonal change, with levels of oxygen and pH rising in summer and decreasing in winter. This pattern is found in all water bodies to a greater or lesser extent, but is most pronounced in highly productive fresh or coastal waters and least evident in oligotrophic (nutrient deficient) rivers.

The reader might well ask how it is that water can hold more than its saturation value. There are several explanations for this. Firstly, since pressure increases with depth, the theoretical saturation value only applies at the surface, and the greater the depth the more oxygen can be held in solution without bubbling out into the atmosphere. Secondly, the solubility of *pure* oxygen in water (i.e. in contact with an atmosphere of pure oxygen) is around 48 mg/l at 15°C compared with the value of 10.07 given in Table 5.

Carbon dioxide

We have already seen that atmospheric gases dissolve in water in proportion to their partial pressure and temperature, as modified by their respective solubilities in water. While air contains almost 21 per cent oxygen and only 0.03 per cent carbon dioxide, the higher solubility of the latter means that its relative concentration in water is proportionally higher. At 0°C the saturation value for carbon dioxide is 1.1 mg/l compared with 14.63 for oxygen, while at 15°C the value becomes 0.6 mg/l compared with 9.8 for oxygen. However, these are theoretical values which are valid for water in contact with the atmosphere in the absence of photosynthesis and respiration. In natural waters these twin processes exert a strong control on the carbon dioxide as well as the oxygen level and, furthermore, since carbon dioxide reacts with water to produce carbonic acid and bicarbonate, the equilibrium between the various components is highly complex.

In eutrophic lakes with abundant phytoplankton and in large lowland rivers with rich growths of aquatic higher plants, photosynthesis in daylight depresses the carbon dioxide while oxygen is substantially increased and, because of its relationship with carbon dioxide, the pH can rise substantially. The reverse takes place at night when carbon dioxide increases, the pH drops and oxygen decreases.

Hardness

Hardness is a term derived from the water supply industry where it has long been in use to describe the suitability of water for certain domestic and industrial purposes. 'Hard' waters contain high concentrations of calcium and magnesium salts such as bicarbonates, sulphates and chlorides which are precipitated from solution by soap and by heating. Thus hard waters can cause severe problems in central-heating systems and in domestic kettles by deposition of lime (mainly calcium and magnesium carbonate) which creates a stonelike deposit inside boilers, pipes and kettles. The washing of clothes in hard water requires larger quantities of soap or detergent than in soft water which has low concentrations of these substances.

Natural hard waters are typically found in areas of calcareous sedimentary rocks such as chalk and limestone, while soft waters occur in areas of basic or metamorphic rocks such as granite and slates.

In the past there have been attempts to classify fresh waters on the basis of hardness because it was thought that this was an important factor determining the distribution of flora and fauna, notably animal species such as molluscs and invertebrates which require calcium for their shells or exoskeletons. It is now recognized that the correlation of species with hard or soft waters is not solely due to the concentration of calcium and magnesium but probably also to a wide range of chemical factors including acidity, alkalinity, bicarbonate, other dissolved salts and available nutrients. Hardness is of specific importance in that it influences the toxicity of many heavy metals such as copper and lead to aquatic organisms, but otherwise is not a particularly useful term for describing the quality of waters in rivers and lakes.

BIOLOGICAL CHARACTERISTICS

It is increasingly acknowledged today that the ultimate test of water quality in the freshwater or marine environment lies in the health of the whole ecosystem of the water body under investigation. A chemical analysis may give a good indication of water quality *at the time of sampling,* but cannot indicate whether a damaging pollutant was present a few hours earlier or whether the dissolved oxygen might have decreased to dangerous concentrations during the previous night. Even regular monitoring at weekly or monthly intervals can fail to detect intermittent conditions which may damage plant or animal species. Moreover, even the best available continuous quality monitoring instruments can only monitor a few parameters, and the high cost of such instruments means that they can only be deployed at a few important locations such as water supply intakes.

The advantages of biological surveys are many but can be summarized as follows:

- Because aquatic organisms live for weeks, months or years, intermittent events such as an accidental spillage of a pesticide, or a short period of acid conditions following an episode of acid rainfall, will have an impact on the ecosystem which can be detected by biological examination for weeks, and sometimes months, or years after the event. For example, trout eggs are sensitive to acidity because highly toxic aluminium may be leached from the soil and kill all the eggs in a headwater stream. Fish counts in later years will show that a particular year-class is missing, indicating that a catastrophic event took place in that year. Similarly, a biological examination downstream from a factory outfall may show a serious depletion of insect species, thus indicating that there may be an intermittent discharge of an illegal substance.

- Only thorough biological surveys can show whether all components of an aquatic ecosystem are present and in a healthy condition. Thus, if most plants and animals are present but there is a deficiency of invertebrates, it may be suspected that a pesticide such as *Permethrin* (specifically toxic to insects but not to fish) may be involved, and further tests can then be carried out to resolve the problem.

- Environmental quality standards (EQSs) which, in effect, ensure that safe levels of toxic substances are set in effluent discharges, depend for their accuracy and effectiveness on a range of biological assay tests in fresh and salt waters. In the UK, the Department of the Environment has set a large number of EQSs under the Dangerous Substances Directive (Table 6).

- Biological monitoring is an essential requirement to ensure that consent conditions for discharges have been correctly set and are achieving the desired objectives in the receiving water.

- Biological surveys are also essential for fisheries management purposes and are valuable for a number of other purposes such as nutrient studies, examining the impact of recreational uses of waters, and the suitability of raw waters for treatment for potable supply.

It is hardly necessary to point out that biological surveys, while valuable, are rarely adequate in isolation and that, for most water management purposes, a combination of physical, chemical and biological information is essential. Populations of plants and animals are constantly responding and interacting with their physical, chemical and biological surroundings, and great care and experience is therefore required in the interpretation of biological information.

Aquatic Communities

During recent years, there has been much emphasis on the presence of 'biological indicators' or 'indicator species' in detecting pollution and in the assessment of water quality. For example, the presence of certain species of algae may indicate acidic conditions while certain insect larvae are only found in well-oxygenated waters. Another way in which indicator species are valuable is the use of species which are known to accumulate toxic chemicals to monitor their environmental distribution. Mussels are particularly useful for monitoring trace metals and micro-organic pollu-

Table 6. Examples of Environmental Quality Standards (EQSs) to protect most sensitive forms of aquatic life (e.g. salmonid fish)

Substance	EQS		Measured as
	Freshwater	Seawater	
Ammonia	15*	21*	A (un-ionized)
Arsenic	50	25	AD
Chromium	5–50	15	AD
Copper	1–28	5	AD
	0.5–12*		
Iron	1000	1000	AD
Lead	4–20	25	AD
		10*	
Nickel	50–200	30	AD
	8–40*	15*	
Tin (inorganic)	25* AT	10*	AD
Zinc	8–125	40	AD
Tributyltin	0.02	0.002	T MAC
Permethrin	0.01	0.01	T 95 percentile
Atrazine and }	2*	2*	AD
Simazine	10*	10*	MAC D
Endosulphan	0.003*	0.003*	AD
	0.3*	–	MAC D
Malathion	0.01*	0.02*	A
	0.5*	0.5*	MAC

Values provided by the National Centre for Environmental Toxicology at WRc plc.

Note: Values are expressed as micrograms per litre (= parts per thousand million by weight). Abbreviations: A = Annual average. D = dissolved. T = total. M = maximum. MAC = maximum allowable concentration.

The toxicity of metals in freshwater is related to the hardness of the water and therefore a range of values is shown, the most stringent standards being required for very soft waters (i.e. containing low concentrations of calcium and magnesium bicarbonate).

*Proposed or revised standard not officially adopted at time of publication.

tants in the marine environment and a routine 'mussel watch' sampling programme is carried out every five years around the UK coastline.

However, for most water quality assessment purposes, it is desirable to study the species diversity of the whole aquatic community or, at least, a major part of it. The reasons for this are best understood by first considering, very briefly, the interrelationships within aquatic communities or *ecosystems*.

Structure of Aquatic Communities

All aquatic communities comprise an assemblage of plants and animals ranging from single-celled organisms such as *Amoeba* or *Euglena* to large plants or animals such as the water lily or a salmon. The whole community is dependent on primary producers such as algae, certain bacteria, and higher plants which contain chlorophyll and are thereby able to utilize the energy in sunlight to produce carbohydrates, fats and proteins from carbon dioxide, water and inorganic nutrients.

The primary producers in lakes, rivers and the marine environment (just as in the terrestrial environment) provide the food (energy input) directly or indirectly for the whole community through what is described as the 'food chain' or, more accurately the 'food web'. Thus the primary producers (algae, bacteria and aquatic plants) are eaten by animals ranging from single-celled protozoa to insect larvae and herbivorous fish, i.e. the primary consumers. These, in turn, form food for the carnivorous species (secondary consumers) including larger invertebrates, amphibia and fish, which, in turn, provide food for larger fish, birds and mammals. This assemblage of species forms what is known as the trophic pyramid (Fig. 5), so called because each succeeding level is smaller in mass than the one below. Not only do the populations (in terms of numbers of individuals) decrease substantially at each trophic level but the *biomass* (weight of biological material) decreases as one moves up. Therefore the biomass of fish in any water body is only a small fraction of the total biomass of all the organisms.

The complex relationships within a food web obviously differ greatly between lakes, rivers and coastal waters, and also between similar water bodies according to a range of factors. One example of a food web is shown in Fig. 6.

Biological Surveys for Water Quality Management

Routine biological surveys are necessary, in conjunction with physico-chemical surveys, to ensure that satisfactory quality objectives are attained and, in particular, that legally binding environmental quality standards (EQSs) under UK or international legislation are being met. Careful design of surveys is essential if they are to provide the necessary information, bearing in mind the staff time which is available. Studies of flora and fauna are time-consuming because of the time necessary for the identifi-

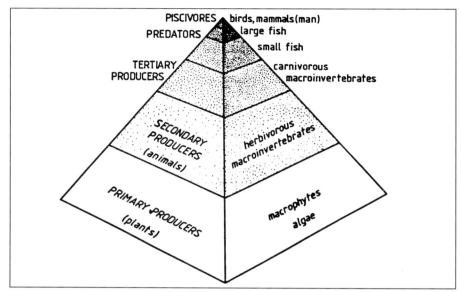

Fig. 5 Diagram of a 'trophic pyramid' indicating relative biomass at each trophic level
(Reproduced with permission from Hellawell, J. M. 1986)

cation and counting of species. A single survey of a river or an estuary may result in weeks of laboratory time. Thus, while it is theoretically desirable to look at the whole ecosystem, in practice it is necessary to limit surveys to specific sectors of the whole aquatic community.

In rivers, it is common practice to sample the macrobenthic fauna at sites which are physically comparable, e.g. riffles where the water flows fairly swiftly over a bed of gravel and sand. In lakes it is usual to examine the phytoplankton and zooplankton while, for fishery management purposes, in both cases, occasional fish population studies are desirable, usually involving electro-fishing and statistical information from the anglers (Plate 2).

In estuaries, studies of the macrobenthic fauna combined with the use of scientific trawling to study the fish species are often used. In coastal waters, transects of the littoral area (intertidal zone) may be used to study the species of animals grazing on

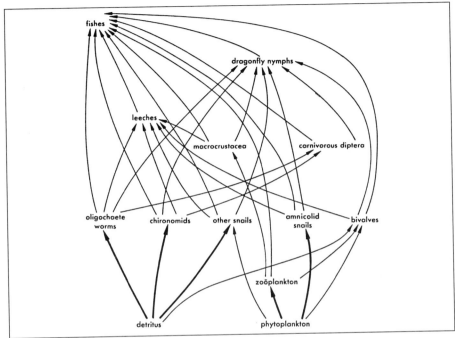

Fig. 6 Simplified food web for organisms in a mature river. (Note that animals at higher levels are not restricted to a single prey species, but feed on organisms from at least two levels)
(Adapted from Carlson, C. A. 1968; reproduced with permission from Russell-Hunter, W. D. 1970)

or sheltering within the seaweeds. In deeper coastal waters, studies of the macro-benthic bottom fauna and plankton surveys will provide good information on the impact of pollutants.

Aquatic Flora and Fauna

It is well recognized that life on earth began in the early oceans and freshwater bodies, and evolved for millions of years before species appeared which were adapted and able to survive on land or in the air. Water is, of course, essential to all life processes, and since most living organisms comprise 90 to 99% water they have had to evolve strategies to retain water and minimize its loss. The populations of water bodies reflect their geography, climate and geological history. For example, freshwaters in the UK contain only modern species of fish, many having disappeared during the ice ages. Waters in the UK contain about 55 species while European waters, also strongly affected by glaciation, have over 100 species.

In contrast, the River Nile, which has been flowing for over 300 million years, has over 300 species of fish, and African freshwaters as a whole have over 2000 species, while in South America the Amazon alone has about 1100 freshwater species. Tropical waters also contain relics of early primitive species, some of which are descendants of the first organisms which began the move from water to a terrestrial existence. In the River Nile today there is the African lung fish which can travel short distances overland between water bodies and survive long dry periods by encystment in the river bed. Also there is an interesting plant, *Fritschiella*, which grows each year on the drying mud banks of the river and is believed to be one of the first land plants. The oceans are, of course, much older than any freshwaters and contain many examples of primitive species, some of which began the move to life on land.

Aquatic Flora

Bacteria

Although invisible to the naked eye, bacteria are universally present in all water bodies, particularly in the sediments of rivers, lakes and estuaries and around particulate organic matter. As already noted, they play a key role in the nitrogen cycle but are also involved in a wide range of other biochemical processes.

The activity of certain bacteria may produce visible effects. For example, in rivers affected by acid waters containing ferrous iron, the iron bacterium *Thiobacillus ferrooxidans* oxidizes the iron to insoluble bright orange deposits of ferric hydroxide. In rivers and estuaries the presence of black muds indicates the activity of sulphur bacteria such as *Desulphovibrio* which thrive in anaerobic conditions and reduce sulphates to hydrogen sulphide, which may be released as gas bubbles. There are other species of sulphur bacteria which oxidize hydrogen sulphide to sulphate. One of these, *Beggiatoa alba*, may form a visible greyish coating on the river bed where there is abundant decomposing organic material such as dead leaves, or in heavily sewage polluted rivers.

Bacteria play the central role in the natural self-purification of waters by decomposing organic matter to its inorganic components which are then recycled. Those involved in the nitrogen cycle have already been mentioned, but there are many specialist species which deserve brief mention.

Lignin, a complex plant material found in algae and higher plants, is slowly degraded by a range of bacterial species under aerobic conditions. It is of interest to note that lignin, a major component of wood, is not easily degraded under anaerobic conditions; therefore wood in bottom muds, especially peaty deposits, may be preserved for long periods.

Cellulose, another plant product, and chitin, which composes the exoskeleton (shell) of insects and other crustacea, are both broken down by specialist chitin and cellulose decomposers under both aerobic and anaerobic conditions.

In addition to these organic compounds, bacteria are capable of breaking down a wide range of chemical compounds, including some pesticides – in fact it is now a requirement for new pesticides to be biodegradable following the earlier experience with organochlorine pesticides such as DDT and Dieldrin which are strongly resistant to decay and persist in the environment for very long periods.

Pathogenic organisms (bacteria and viruses)
Pathogens are of particular concern in surface waters which are used as a source of potable supply or for bathing. As far as potable supplies are concerned, modern water treatment processes, including disinfection, ensure that sources are well protected and that any pathogens present are removed by filtration or killed.

However, the treatment of urban wastewaters does not remove pathogens, therefore rivers and coastal waters are liable to contain enteroviruses and bacterial pathogens such as *Salmonella* and *Shigella* *(see Bathing Water Directive in Appendix).*

Cyanobacteria
This is an important group of plants which, until recently, were classified as algae (the blue-green algae). Cyanobacteria are at times a dominant part of the phytoplankton in lakes and reservoirs, often appearing in the late summer or autumn after earlier growths of diatoms and green algae. They tend to form dense surface blooms which form suddenly because their cells are buoyant, and on calm windless days they rise to the surface and drift with any currents to form massive suspensions which may be up to one metre in depth. They are of major health concern because some species produce highly toxic secretions which cause skin irritation and serious stomach upsets if consumed. Cattle and pet animals have been known to die after drinking water from affected lakes. Cyanobacteria are not significant from a water quality standpoint in rivers or tidal waters. The main genera found in British waters are *Microcystis*, *Anabaena*, *Nostoc*, *Aphanizomenon* and *Rivularia*.

Algae

Phytoplankton monitoring provides the best biological assessment of water quality in lakes and reservoirs. Such waters are characterized by seasonal successions of species which are often remarkably similar from year to year, so that any sustained trends towards changes in species or productivity may indicate changing water quality. During the winter months, algal activity is minimal because of low temperatures and low light intensity. In the spring months, diatoms often appear first and reach peak growths in May/June – often limited by the supply of silicate. They may die off in early summer and be followed by green planktonic algae. Recovery of silicate levels may produce a second, smaller crop of diatoms in the early autumn and, often at this time, a large bloom of cyanobacteria (formerly known as blue-green algae). Estimates of crops can be made by counts of cell numbers per unit volume and by the measurement of chlorophyll concentration. Many lakes and reservoirs are subject to stress because of increasing nutrients and phytoplankton density, and therefore surveys of this kind are an essential prerequisite to the design of policies to limit nutrient enrichment.

In coastal waters, surveys of phytoplankton productivity are also valuable in the study of eutrophication, especially in the vicinity of known inputs of nutrients such as rivers and sewage outfalls. Monitoring of marine phytoplankton is currently undertaken by chlorophyll measurement, which is now required in vulnerable waters under the EC Urban Wastewater Treatment Directive. Seaweeds, notably the brown algae (Phaeophyceae) and red algae (Rhodophyceae), comprise the dominant flora of the inshore coastal waters and are important providers of shelter and food for a wide range of marine animals. However, although studies of these communities in relation to pollution have been carried out, they are not currently used for routine monitoring.

Algae are sensitive to many environmental factors, notably the pH (acidity/alkalinity) of the water. This observation has provided a well-proven research technique for reconstructing past pH values in lakes, extending, in some cases, for hundreds of years. Certain diatom species are known to have specific pH requirements so that the species composition can be used to estimate the acidity of lake waters within fairly close limits. Because of their hard silica shells, diatoms are extremely well preserved in lake sediments, and the diatom analysis of sediment cores provides an excellent method of reconstructing their historical pH values. Studies from many UK lakes in areas which are geologically sensitive to acid deposition have shown that there has been a significant decrease in pH since about 1800 AD (Fig. 7).

Higher plants

A range of plants, including liverworts, mosses and flowering plants, are found to grow in streams and standing waters. Their presence and distribution is related to factors

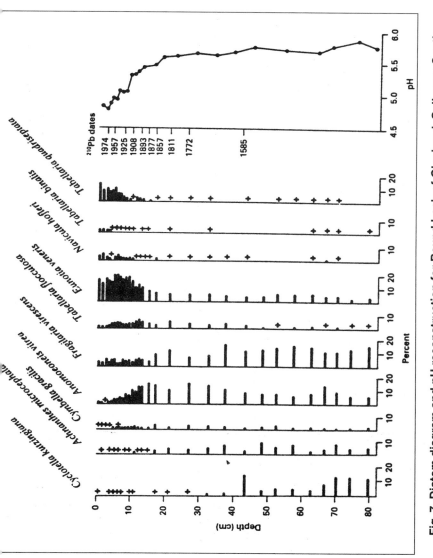

Fig. 7 Diatom diagram and pH reconstruction for Round Loch of Glenhead, Galloway, South Wear, Scotland

(From Flower & Battarbee, 1983; reproduced with permission from Mellanby, K. 1988)

which include altitude, temperature, flow rate, nature of the lake or stream bed and water quality. In the upper reaches of streams, for example, mosses such as *Fontinalis antipyretica* are often abundant in clean water but may be replaced by the alga *Cladophora glomerata* (blanket weed) in waters enriched by nutrients derived from sewage effluents or agricultural fertilizers. The abundance of higher plants tends to increase with nutrient enrichment and also in slow-flowing sections where the build-up of sediment allows seeds to take root. As more plants become established this further reduces current velocity, thus increasing sedimentation and the establishment of yet more rooted vegetation. If this happens in midstream, the build-up may lead eventually to the formation of islands. Where rooted plants become abundant, their presence can itself cause quality problems such as severe deoxygenation by night and high alkalinity by day. In hot summer weather, both of these conditions can sometimes result in the death of fish life. Surveys of aquatic vegetation are not generally used in routine water quality studies. Nevertheless, biologists carrying out surveys of aquatic fauna find that it is of value to record the plant species which provide shelter and food for many species of invertebrates and young stages of fish.

Aquatic Fauna

As seen earlier, animals comprise the primary and secondary consumers in natural ecosystems. The primary consumers include herbivorous organisms which graze on algae and higher plants and the detritus feeders which feed on the decomposing remains of plants. Herbivores in the aquatic environment range enormously in size from microscopic protozoa to herbivorous fish. Secondary consumers include a great variety of carnivorous animals from insects to fish, while at the top of the 'pyramid' are fish-eating birds and mammals (Fig. 5). Animals which feed on detritus from decomposing animal remains are also included in this category.

The biological classification of rivers in the UK is almost entirely based on the assemblages of animals known as 'macrobenthic invertebrates' – in other words the larger bottom-living animals which can easily be collected by means of a hand net. In order to obtain comparable samples, the net is taken across the width of the river, usually in a moderately fast flowing 'riffle' section for a standard time, e.g. 30–60 seconds while wading and kicking or otherwise disturbing the stones and silt to release the animals which are carried into the net by the current (Plate 3). The net may be emptied into a white enamel tray for immediate examination on the river bank or, more usually, emptied into a sample bottle to which a preservative such as alcohol or formaldehyde is added for later examination under a microscope in the laboratory (Plates 4, 5, 6).

The animals may include flatworms, worms, leeches, snails, limpets, shrimps, waterlice, water mites and spiders, and a large array of insect larvae such as stoneflies, mayflies, caddisflies, dragonflies, midge flies, beetles and bugs (Fig. 8).

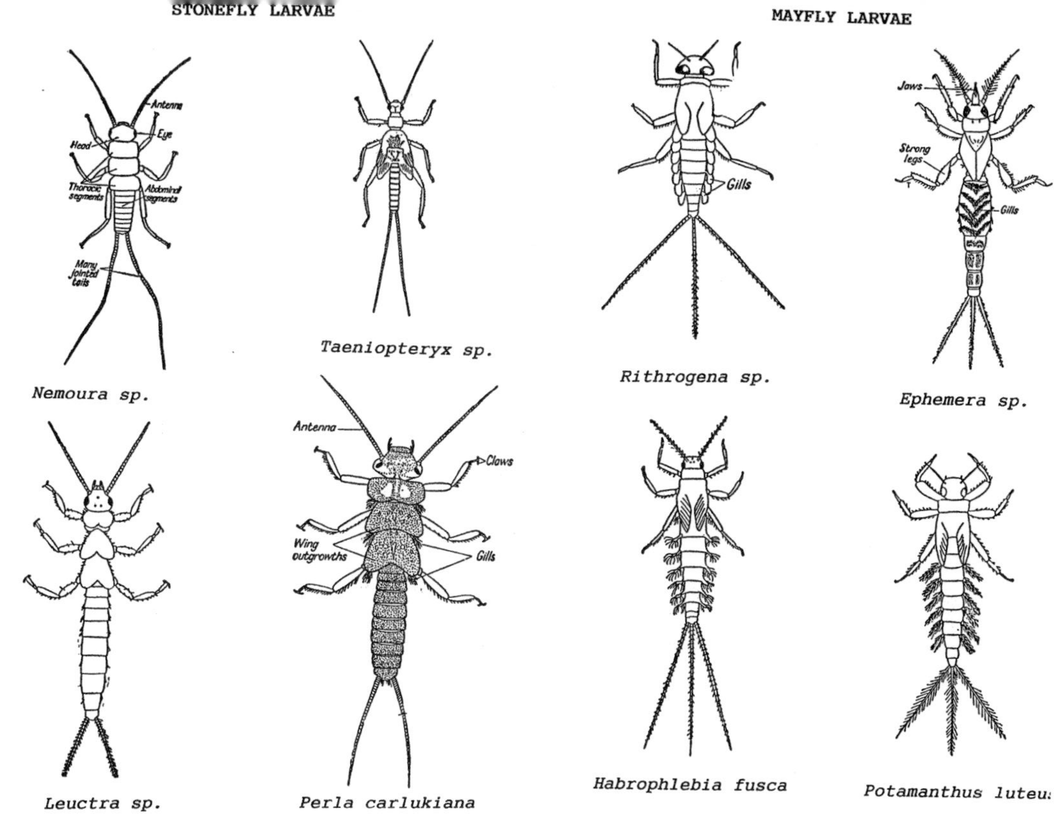

Fig. 8 Biological indicators of pollution: these stonefly and mayfly larvae are examples of clean water fauna

(Adapted from Mellanby, H. 1953; reproduced with permission from Chapman and Hall)

Most of the insects mentioned here spend a large part of the year as larvae feeding in the river bed or on aquatic vegetation, and emerge as adult flies in spring, summer and autumn. The adults of many species only live for a few days or weeks, their main function being to mate and lay eggs in rivers, ponds or lakes, thus distributing the species more widely and ensuring survival into the following year.

Each of the hundreds of species of invertebrates is adapted to live within a certain range of temperature, water velocity, habitat and chemical water quality conditions such as pH, dissolved-oxygen level, calcium concentration and nutrient availability. Thus, the fauna of an acid moorland stream will be very different from that of an upland chalk stream, while in both cases differences in current velocity, temperature range and levels of nitrate or phosphate will produce changes in species composition. Oligotrophic waters (deficient in nutrients) tend to have sparse populations and moderate species diversity except in acid waters where there will be a reduction in species. Moderately eutrophic waters (rich in nutrients) are likely to have a fairly high population density and also a fairly high diversity of species, while highly eutrophic (very nutrient rich) waters may have exceptionally high numbers of animals but a more limited range of species. In addition to physico-chemical factors, the presence of particular species may also be related to the other species of plants and animals which are present.

Routine biological surveys of lakes (and reservoirs) for water quality purposes are generally confined to the plankton and, in conjunction with phytoplankton mentioned earlier, are concerned with population studies of water fleas (Cladocera) such as *Daphnia* and *Bosmina*, Ostracods such as *Cypris*, and Copepods such as *Cyclops* (Fig. 9).

In estuaries and coastal waters, biological assessment of water quality is largely dependent, as in rivers, on surveys of the bottom-living animals living on and within the estuarial and coastal sediments. Routine surveys in UK waters, to study the impact of coastal outfalls or the pollution status of estuaries, have mainly used macrobenthic species collected by grabs. The main groups of animals comprise Polychaete worms, Crustacea, Molluscs, and Echinoderms (sea urchins). In estuaries, variable inputs of freshwater from inflowing rivers can cause large population changes, and caution is needed in interpreting the results, which should be based on samples taken at various seasons over a number of years if, for example, recovery from pollution is being assessed. (Fig. 10).

IMPACT OF POLLUTANTS

Pollution has been defined in many ways, but the following definition is now widely accepted:

'Pollution of the aquatic environment means the introduction by man, directly or indirectly, of substances or energy which cause such deleterious effects as (1) harm to living resources, (2) hazards to human health, (3) hindrance to aquatic activities, (4) impairment of water quality for use in agricultural, industrial and economic activities, and (5) reduction of amenities.'

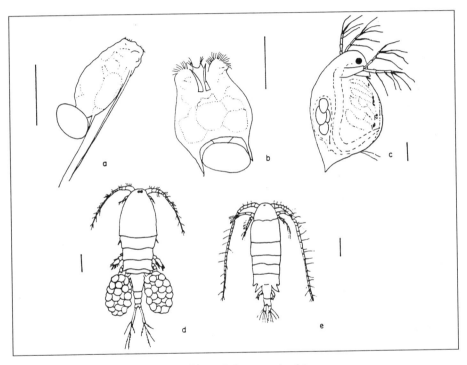

Fig. 9 Some lake zooplankton
Rotifers: (a) *Filina*, (b) *Brachionus*
Cladocera: (c) *Daphnia*
Copepods: (d) *Cyclops*, (e) *Diaptomus*
(scale lines indicate a length of 0.1 mm)
(Reproduced from Moss, B., 1980, with permission from Blackwell Scientific Publications)

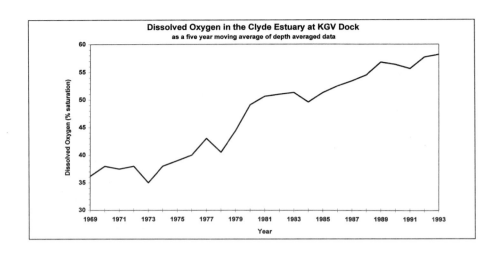

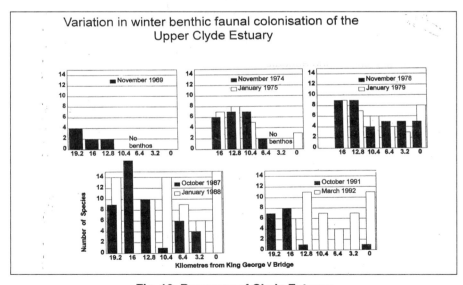

Fig. 10 Recovery of Clyde Estuary
(Reproduced by courtesy of the Clyde River Purification Board)

It is important to note that, under this definition, pollution is man-made and, therefore, that natural inputs of similar substances or energy are not included. Pollution is therefore an addition made by man to naturally occurring biogeochemical inputs.

Pollutants can be classified into groups according to their composition and effects as follows:

Organic Wastes

Organic wastes include a wide range of effluents such as domestic sewage, effluents from paper pulp mills and the liquid wastes from food-processing industries such as abattoirs, fruit and vegetable canneries, breweries, and distilleries. Another potentially serious source of organic pollution is the petroleum industry, usually through accidental spillages of crude oil, heating oils, diesel, paraffin or petrol. However, there is little doubt that the biggest source of organic pollution to inland and coastal waters is domestic sewage, particularly in developed countries where every house, shop, office and factory has a piped supply of water which is subsequently discharged to the sewerage system, carrying with it a vast load of waste organic matter.

Organic matter, whatever its source, comprises cellular material from plants or animals which is naturally decomposed by bacteria utilizing the dissolved oxygen in the receiving water. The capacity of a wastewater to consume oxygen is known as its *biochemical oxygen demand* (BOD). Raw, untreated sewage and most organic wastes have a high BOD and cannot be allowed untreated into receiving waters because they deoxygenate the water causing fish and invertebrates to die, giving rise to objectionable smells and appearance. Sewage and many other such wastes usually contain suspended organic solids which may settle out in the vicinity of the outfall in the form of a black sludge which obliterates most of the bottom-living animals.

Sewage contains ammonia which is highly toxic to fish and also detergent residues which make fish more sensitive to ammonia and low dissolved-oxygen concentrations. Sewage, especially in industrial communities, often contains many other pollutants, and the objective of treatment is to ensure that the organic load and other pollutants are reduced to concentrations which are harmless to aquatic ecosystems. Petroleum products, even when discharged in very small amounts, cause unsightly surface slicks and are also toxic in varying degrees. Even in low concentrations they may cause tainting of fish.

BOD is measured by incubation of samples of water in the dark at 20°C for 5 days and determining the consumption of oxygen. For samples with high BOD levels, dilution with distilled water is necessary to avoid total consumption of the dissolved oxygen.

The main effects of organic wastes on streams are shown diagrammatically in Fig. 11.

Nutrients

The treatment of sewage and some other organic wastes tends to produce effluents rich in nitrate and phosphate which, in turn, may cause excessive growths of plants and animals. Nutrient enrichment or 'eutrophication' is now recognized as a potentially serious problem in inland and coastal waters. Nutrient removal is now carried out at only a few effluent treatment plants in the UK but is likely to become much more widespread under the EC Urban Waste Water Treatment Directive (see Appendix).

Inert Solids

It might be thought that inert solids such as sand, clay and other mineral particles would be relatively harmless in rivers and coastal waters. However, it has been found that continuous discharges of even relatively small amounts of fine solids can harm the breeding grounds of fish in rivers by filling the voids in gravel beds, thus preventing oxygenated water reaching the eggs of fish and also eliminating those insect species which seek shelter within the gravel of lakes and streams. Moreover, fine solids in suspension cause unsightly slicks and may reduce light penetration. + irritate fish gills.

Metals

Most metals such as aluminium, arsenic, cadmium, chromium, copper, lead, mercury, nickel and zinc are extremely toxic to aquatic organisms, particularly when they are in the ionic form. Moreover, since metals may be precipitated into sediments near the outfalls, much higher concentrations may be built up than in the water above. A further problem is that some organisms accumulate metals, and these may further build up in the larger animals at the top of the food web. These processes, respectively known as *bioaccumulation* and *biomagnification* must be taken into account in determining EQSs.

Methods for the determination of metals require scrupulously clean techniques because of the low levels and the risk of contamination. Atomic absorption with flame atomization is widely used but will require pre-concentration for levels below about 50 µg/l. These methods are increasingly being replaced by atomic absorption with electrothermal atomization and now by inductively coupled plasma atomic emission spectrometry.

Pesticides

In recent decades, pesticides have proved to be a major environmental problem with about 10 000 different formulations on the market comprising insecticides, herbicides, molluscicides and fungicides. Many of these have long lifetimes in the environment

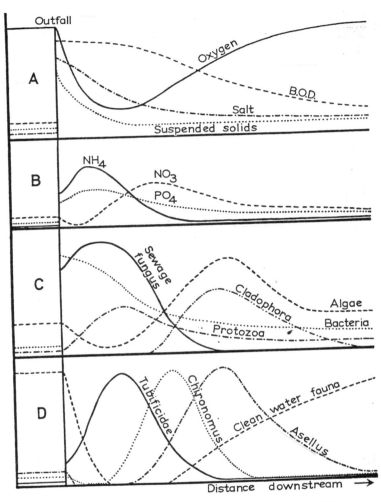

Fig. 11 Diagrammatic presentation of effects of an organic effluent on a river and changes on passing downstream from outfall. (A and B physical and chemical changes, C changes in micro-organisms, D changes in larger animals)
(Reproduced with permission from Hynes, H. B. N. 1960)

and their breakdown products may also be highly toxic. The problems are compounded by the fact that, like metals, they are subject to bioaccumulation and biomagnification, and they are often toxic at levels which are close to or below the levels of accurate determination. They are extremely difficult to control because the main sources are diffuse from agricultural land or forests and many are dispersed on a global scale by evaporation into the atmosphere and redistributed in rainfall.

Analysis for pesticide residues in water requires sophisticated methods such as gas chromatography combined with mass spectrometry. For pesticides which cannot be extracted from water and which cannot withstand heating, high-pressure liquid chromatography is being increasingly used.

Waste Heat

Thermal pollution is mostly caused by cooling water discharges from power stations and some factories. By raising the ambient temperature of the receiving water the solubility of gases such as oxygen is decreased and the biochemical processes which consume oxygen are speeded up – a generally noted rule being that the rate of metabolic processes such as respiration is doubled for each 10 degrees rise in temperature. The effect of thermal pollution therefore can be very similar to organic pollution. As far as the impact on biota is concerned, raised temperatures may slightly speed up the life cycles of invertebrates and may increase the risk of fish deaths at low flow in summer when rivers are already at a high temperature. However, in general these effects are not great, provided that adequate controls are imposed by the regulatory authority.

Pathogens

Organic wastes, particularly domestic sewage, usually contain large numbers of micro-organisms and these may include pathogens such as viruses, bacteria and protozoa. The organisms responsible for highly infectious waterborne diseases such as cholera and typhoid are not normally found in sewage in developed countries because there are no longer any infected individuals in the countries to provide a source. However, the cholera vibrio and typhoid bacillus are often present in effluents and rivers in developing countries and, of course, were responsible for many epidemics in Europe in the nineteenth and early twentieth century. Disinfection of effluents is not generally practised in the UK, and sewage effluents may contain such pathogens as species of *Salmonella*, Polio virus, Hepatitis virus and bacteria such as *Leptospira* which is carried by sewer rats and causes Weil's disease. The most sensitive test for sewage pollution is that for *Escherichia coli* but, under the EC Bathing Water Directive (see Appendix), it is now a requirement to test for total coliforms, *E. coli*, and, if their presence is suspected, enteroviruses and *Salmonella* in identified bathing waters. These tests are routinely carried out at about 460 bathing waters around the UK coastline during 20 weeks of the bathing season.

Acid Deposition

It is believed that rainwater has always been acidic with a pH around 5.6 due mainly to small natural levels of carbon dioxide and sulphur dioxide in the atmosphere. The burning of fossil fuels in power stations and factories in industrial countries has greatly increased the amount of sulphur dioxide while oxides of nitrogen, mainly from motor vehicles, are another source of acid deposition.

The effect of acid deposition on inland waters is worse in areas of coniferous forest. The mechanism here is that the pine needles collect dry, particulate sulphur which, during wet periods, is washed into the soil and then to the watercourses. The subsequent impact depends almost entirely on the geology of the rocks and soils. In areas where the rocks are rich in calcium carbonate (chalk and limestone) the acidity is completely neutralized, but where the rain falls on granites and slates containing very little calcium, the acidity cannot be neutralized and surface waters may become very acid during episodes of acid rain. Acidity has serious effects on the flora and fauna. Below pH 5.5 aluminium is leached from soils and, depending on its form, may kill fish, especially the eggs and larval stages. Inorganic aluminium is the most toxic but it can form chemical complexes with organic acids, fluorides and silicates which reduce the bioavailability of the aluminium and reduce its toxicity. Invertebrates are less affected by aluminium but the acidity alone greatly reduces species diversity.

Acidification can be of concern where waters are used for drinking because of increases in aluminium, copper, nickel and zinc, particularly in groundwaters.

Critical Loads

The critical loads concept has been developed within the United Nations Economic Commission for Europe (UNECE) as a method of quantifying the sensitivity of ecosystems to pollutants. Its first application has been to the problem of acidification, and this has been welcomed by the UK Government.

A 'critical load' is defined as a load of pollutant which a particular ecosystem can tolerate without adverse effects. Therefore in aquatic systems the critical load is set at a level which will ensure a water quality suitable for the maintenance of healthy fish and the ecosystem which supports them. The critical loads approach has been used to prepare a map of the UK using a 10 x 10 km grid which shows (for each square) the estimated critical load, taking into account the geology, land use and soil type. This approach has been used to estimate the reductions required under the second stage of the UNECE protocol on nitrogen oxide emissions from 1994. An existing UNECE protocol has required a 30% reduction in sulphur dioxide emissions by 1993 as compared with 1980.

WATER QUALITY ASSESSMENT

An assessment of water quality is required by the regulatory authorities listed on page 55 for many purposes, including the design and enforcement of consent conditions, compliance with a large number of national and international EQSs, the suitability of water for abstraction for potable, agricultural and industrial purposes, and the provision of annual and quinquennial water quality reports to the Government. This information is largely derived from routine monitoring supplemented when required by special surveys.

Former National Rivers Authority (NRA) and River Purification Authority (RPA) Monitoring Systems (see p. 55)

Each NRA Region and each River Purification Board (RPB) had a long established network of sampling stations on its rivers. These stations are mostly situated at accessible key points to determine the impact of major discharges and tributaries. Samples are usually taken in mid-stream at mid-depth at monthly intervals and analysed for a wide range of physico-chemical parameters. At some of the more important points, samples may be taken weekly, while in upland areas remote from point sources samples may be taken quarterly (Plate 7).

Sampling stations for biological examination are situated as close as possible to the chemical survey stations but need to be accessible for wading with a hand net or for grab sampling from small boats.

In estuaries such as the Clyde and Thames, and in coastal waters, larger sea-going boats (Plate 8) are normally required, and routine water quality surveys are made several times per year, usually on a central line down the estuary and at a range of depths. In coastal waters, sampling may follow transects from shallow inshore stations to deeper stations offshore, also at various depths, to determine the impact of long outfalls and general trends in water quality (Plate 9). For biological purposes, samples in both estuaries and coastal waters are taken by grabs lowered by winch to the bottom (Plate 10). These can be supplemented by divers equipped to take samples *in situ* and by the use of underwater cameras aided by closed-circuit television to photograph the bottom fauna using flash illumination. Coastal waters of England and Wales are normally sampled quarterly at 15-km intervals.

Harmonized Monitoring Scheme (HMS)

In 1974, the DoE established the UK Harmonized Monitoring Scheme which comprises over 200 stations at the tidal limits of major rivers and above the junctions of important tributaries. Analyses are carried out for about 25 determinands regularly and over 100 altogether. The analyses are carried out by specified methods with analytical quality control to ensure consistently accurate results. The scheme had two main purposes, (i) to estimate the loads of pollutants entering UK coastal waters and (ii) to identify long-term trends in river water quality. The scheme is still in operation but, unfortunately, few data have so far been published.

International systems

The Harmonized Monitoring Scheme did, however, provide an early model for later international schemes, notably the Global Environmental Monitoring System (GEMS) and a smaller European system set up under the Exchange of Information Directive. A proportion of UK HMS stations form part of these networks which are also subject to analytical quality control.

In the marine environment, information from a proportion of UK monitoring sites must be reported to the Joint Monitoring Group, the North Sea Ministerial Conference, and to the Oslo and Paris Commissions.

Water Quality Modelling

Computer-based mathematical models are an important tool in water quality management and are widely used in (a) predicting the impact of pollutants on water quality, (b) comparing alternative strategies in environmental management, and (c) designing consent standards to achieve specified objectives.

The majority of models used in rivers and estuaries to date have been used to predict oxygen levels using such data as river flow, dilution, BOD, dissolved oxygen and temperature. Models are also used to predict levels of ammonia and nitrate along the length of rivers at various flows and under critical conditions of minimum flow and high temperature.

In estuaries, more complex models are used in which freshwater inputs, benthic oxygen consumption, nitrification, etc. are also required, together with tidal water movements and salinity data. Such models, which can predict the dissolved-oxygen concentrations through the length and depth of the estuary, need to be validated by comparing predicted values with actual data for a range of conditions.

Models are now routinely used in designing consent standards for new effluent treatment plants, in the prediction of flood levels and in estimating runoff from ungauged river catchments. They are also used in developing strategies for controlling a series of polluting discharges along the length of a river and for modelling the dispersion of pollutants from long outfalls into deep coastal water.

Some of the most complex models used in water quality management have been multi-layered, multi-compartmented models of lake ecosystems using climatic, chemical, hydrological and biological data to predict the behaviour of plankton communities such as productivity of phytoplankton and the cycling of phosphorus and nitrogen in thermally stratified systems. Also, in acidification control studies, critical loads data (referred to earlier) have been used to predict the effect of differing percentage reductions in airborne emissions of sulphur dioxide over the whole of the UK.

Automated Monitoring Systems

Automatic water quality monitoring stations are expensive to set up and maintain, and are therefore only used where there is good justification for the expense. Automatic stations in rivers are in widespread use upstream from water supply intakes to give early warning of adverse water quality. However, these are limited in the number of determinands which can be monitored continuously. The main ones are oxygen, ammonia, pH, conductivity and temperature. In general they are not capable of detecting trace levels of organic micropollutants such as solvents or pesticides. However, such instruments are valuable in that they can be telemetered to control centres and thus provide a continuous watch on water quality. The development of specific-ion electrodes should enable a wider range of low-level determinands to be monitored.

Automatic floating water quality monitoring stations mounted on buoys are now being used in the marine environment. The data are stored on a data logger and can also be transmitted to a land station by radio signal.

QUALITY CONTROL

Effective water quality management can only be achieved by means of procedures which are reliable and measurements which are accurate within specified limits of error. For many years, checks on accuracy were limited to chemical laboratory analytical procedures by means of specified procedures and inter-laboratory comparison using reference samples from a central control laboratory.

Today it is standard practice to apply quality control procedures to all aspects of an organisation's work – administration, inspection, sampling, analysis, data processing and communication. All the regulatory bodies have obtained, or are now seeking, accreditation under national quality control standards, e.g. NAMAS M10 for laboratory procedures, or international standards such as BSEN ISO 9002 for other functions. Accreditation under these standards is a lengthy process involving the preparation of manuals which set out all the procedures to be followed, the training of staff and regular auditing by the accreditation agencies.

3. LEGISLATION AND MANAGEMENT OF WATER QUALITY

UK AND EC LEGISLATION

In the UK, because of different legal systems, there has been generally separate legislation for England and Wales, for Scotland and for Northern Ireland. In 1974 the *Control of Pollution Act Part II (COPA II)* repealed most of the relevant preceding legislation and provided (for the first time since 1876) in the one Act all the necessary powers for water pollution control. It continued the offence provisions of the Rivers (Prevention of Pollution) Acts whereby it is a criminal offence to discharge (a) any poisonous, noxious or polluting matter to controlled waters and (b) any sewage or trade effluent without the consent of the water authority. It also brought under control, for the first time, all discharges to coastal waters and groundwaters and provided for the implementation in the UK of EC legislation.

A new feature was the introduction of public registers to make almost all the work of the regulatory bodies open to public scrutiny. The Act also provided for public involvement in the granting of consents for significant discharges by requiring that applications for consent be advertised in relevant newspapers, with provision for public objections with an appeal procedure. It is worth noting that, because of the different legal systems and the different regulatory authorities, this Act (COPA II) and subsequent legislation contain various clauses which adapt the legislation to each system. Essentially, the same laws apply throughout the UK with, however, some important exceptions. Thus, in England and Wales the National Rivers Authority had powers to control abstraction and for flood prevention, whereas the River Purification Authorities (RPAs) in Scotland did not have these powers.

In England and Wales *COPA II* was revised and incorporated in the *Water Act 1989*, which also created the National Rivers Authority for England and Wales. However, in Scotland, *COPA II* continued almost unchanged as s.23 of the *Water Act 1989*.

The *Environmental Protection Act 1990* introduced a new procedure known as 'integrated pollution control' (IPC) whereby emissions to air, water or land from large industries with complex, potentially hazardous products are controlled under single authorizations instead of separate consents for each medium.

Insofar as water quality is concerned, control is now largely determined by EC directives and decisions. These have been unanimously approved by the member states as part of a series of Environmental Action Programmes. The first of these Programmes adopted in 1973 had three aims: (i) to reduce pollution and nuisances; (ii) to husband natural resources and the balance of ecosystems, and (iii) to improve the quality of life and working conditions. The 5th Environmental Action Programme was unanimously adopted by the 12 member states in 1992 and sets new targets for the period to 2000 AD. It seeks to strengthen and review existing controls where necessary, and to increase cooperation with industry by voluntary measures to reduce environmental impact. This is a major change of direction from the strictly regulatory approach, but it remains to be seen whether it will prove effective. There is no doubt that protection of the environment is now a major public concern, and a growing number of companies are keen to demonstrate that they are 'environmentally friendly'.

A list of the most important pieces of European Community legislation is given in the Appendix.

INTERNATIONAL AGREEMENTS

The UK has agreed to implement a number of international treaties and other agreements concerned with the marine environment, such as the Oslo and Paris Conventions (concerned respectively with pollution from dumping at sea and from pipelines to the sea) and the decisions of the North Sea Ministerial Conferences of 1987, 1990 and 1995.

Since these agreements were made, concern for the environment has finally reached the global level. At the United Nations Conference on Environment and Development (UNCED) held in Rio de Janeiro in 1992 (the 'Earth Summit') the following decisions were made:

- Agenda 21: a world-wide programme of action for the achievement of sustainable development.
- Convention on biodiversity: an international agreement to protect the diversity of living species and habitats.
- Convention on climate change: a framework for action to reduce the risks of global warming.

- Sustainable forestry: a statement of principles for management, conservation and sustainable development of the world's forests.

Over 140 member states were represented and each state is expected to prepare its own strategy and action plans to implement the above agreements. The UK Government published its strategies and action plans in January 1994 and played a leading role within the European Community in developing the 5th Environmental Action Programme referred to above. Under the strategy for sustainable development, the UK Government has outlined proposals to improve the quality of surface and groundwaters, including measures to encourage environmentally friendly farming by minimizing the use of pesticides and the impact of agricultural wastes.

REGULATORY AUTHORITIES

Until 31 March 1996, water quality control in rivers, lakes, groundwaters, estuaries and coastal waters was the responsibility *inter alia* of the following bodies:

(1) The National Rivers Authority (NRA) which covered the whole of England and Wales and was subdivided into eight Regions based on river catchments.
(2) The River Purification Authorities (RPAs) in Scotland comprising seven River Purification Boards (RPBs) and the three Islands Councils (Orkney, Shetland and the Western Isles) which were all-purpose authorities and therefore also responsible for integrated pollution control (IPC).
(3) The Department of Environment for Northern Ireland which was also responsible for IPC.
(4) Her Majesty's Inspectorate of Pollution (HMIP) which was responsible for IPC in England and Wales.
(5) Her Majesty's Industrial Pollution Inspectorate (HMIPI) which was responsible jointly with the RPAs for IPC in Scotland.

However, on 1 April 1996, under the Environment Act 1995, these organisations were abolished and their powers transferred to newly created bodies, i.e. the Environment Agency (EA) for England and Wales, the Scottish Environment Protection Agency (SEPA) for Scotland and, in Northern Ireland, the Environment and Heritage Agency. These agencies are responsible for all aspects of environmental pollution control.

CONSENT SYSTEM

The consent system, which was first introduced in 1951, is the principal mechanism for controlling water quality in the UK. Basically it is very simple: firstly, it is a criminal offence to discharge any poisonous, noxious or polluting matter to controlled waters and, secondly, it is also a criminal offence to discharge any sewage or trade effluent without the consent of the regulatory body. Consent can be refused, but not 'unreasonably'. Consent is normally granted but with a range of conditions designed to protect the receiving water. The maximum penalty on summary conviction is a fine of £20 000 but, in the event of trial by jury, the fine is unlimited and a term of imprisonment may be imposed. Breach of consent conditions is also a criminal offence which is subject to the same penalties.

The setting of consent conditions is carefully designed to protect each receiving watercourse under the worst conditions of dilution and dispersal. In rivers, consents are designed to protect the water at low summer flows, usually 95 or 97% exceedance flow, i.e. that flow which will be exceeded for 95 or 97% of the time. There are several safety factors in these calculations:

(i) The EQSs for each pollutant are set at concentrations which are at least ten times lower than the lowest concentration at which adverse effects have been observed;
(ii) Dischargers are not normally allowed to take up the whole capacity of the water to absorb pollution;
(iii) Dilution is also based on the maximum permitted flow of the discharge, and for most of the time there will be a smaller volume; and
(iv) In rivers the flow will be higher for most of the time.

There is a further safeguard in that, after the elapse of two years, the consent conditions can be reviewed and made more stringent if monitoring shows any adverse effects, or relaxed if this is justified. However, experience shows that it is only rarely that consents set in recent years need tightening. Of course, if there is any change in the volume or composition of the discharge, the regulator can review it immediately.

In lakes, estuaries and coastal waters, dispersion and dilution are inherently more difficult to calculate, and hydrographic surveys coupled with computer simulation are increasingly used. Often the location and length of outfall are dictated by the need to obtain adequate initial dilution within a small mixing zone around the point of discharge. On long sea outfalls, the degree of dilution can be increased by the incorporation of a series of diffusers along the effluent pipe.

The conditions in each consent cannot be calculated in isolation but must take account of other nearby discharges. The consent should be drawn up so that (a) there is no significant deterioration of water quality in terms of toxicity, bioaccumulation or eutrophication, (b) all relevant EQSs are met, and (c) the discharge causes no adverse annual visual appearance (e.g. slick or discoloration), no smell and no tainting of fish.

When the consent comes into force the discharge is monitored to ensure that the conditions are met. The receiving water is also monitored chemically and biologically to ensure that the consent has been correctly designed and that the environmental quality objectives are fully met.

INTEGRATED POLLUTION CONTROL

'Integrated pollution control' (IPC) is a concept which was first proposed by the Royal Commission on Environmental Pollution (RCEP) and enacted in the Environmental Protection Act, 1990. The aim of IPC is twofold, i.e. (i) to allow regulators to consider all environmental pathways when licensing emissions to air, water and land, and (ii) to simplify procedures for large industries which have emissions to more than one medium by providing for a single application and a single authorization for all the discharges from a particular process. Although already in operation, the full benefits of IPC should be realized with the formation of the environmental agencies.

STATUTORY WATER QUALITY OBJECTIVES (SWQOs)

Until March 1996, the NRA and RPAs based their control policies on non-statutory quality objectives which have provided the basis for setting consent conditions related to water use. However, under the provisions of the Water Resources Act, 1991, a more formalized system should soon come into use whereby the Secretary of State for the Environment (or Secretary of State for Scotland or Northern Ireland as appropriate) will (a) prescribe, in Regulations, a system of classifying controlled waters under the scheme, and (b) periodically serve notice on the regulatory authority specifying a stretch of water and a date by which the water should comply with one or more prescribed classifications.

The specific uses which the Environment Agency envisages as being protected by SWQOs include such possibilities as basic amenity, general or special ecosystems, coarse fishery, salmon fishery and (in estuaries and coastal waters) migratory fishery, commercial fishery, shellfish waters, water contact sports, etc.

While the introduction of SWQOs is imminent in England and Wales, their introduction in Scotland is still under consideration and is even considered as a barrier to progress where the long-term objective of the RPBs has always been to achieve salmonid river status.

CATCHMENT MANAGEMENT PLANS

The RCEP recently stressed the need for catchment management planning, pointing out that the whole of a drainage basin (or catchment) must be taken into consideration for effective water quality management. This should be self-evident when it is recognized that most forms of land use have implications for water quality. The most obvious example is agriculture ,where fertilizers and pesticides inevitably drain to the rivers. Forestry, mining, urban development and many other activities impinge on water quality and therefore catchment management plans (CMPs), which are in their infancy at present, will increasingly feature in long-term planning of environmental water quality. The NRA produced a number of CMPs and more are being developed by the Environment Agency.

4. ENVIRONMENTAL QUALITY STANDARDS AND WATER QUALITY CLASSIFICATION

ENVIRONMENTAL QUALITY STANDARDS

In the UK, it has long been the tradition to set consent conditions in order to maintain a desired quality in the receiving water. This tradition stems from the 8th Report of the 1898 Royal Commission on Sewage Disposal which proposed the famous 'Royal Commission standard' for sewage effluents, which was subsequently used all over the world, often as a fixed standard although that was not the intention of the Commission. The 20:30 standard (20 mg/l BOD and 30 mg/l for SS) was intended for rivers, on the assumption that the effluent would be diluted with at least eight volumes of water having a BOD not exceeding 2 mg/l. If greater dilution was available, a more relaxed figure could be used, and if less than 8:1 a more stringent one. Accordingly, standards of 15:20, 10:15 or 5:8 for BOD and SS are commonly used.

For many years, the early river boards and river authorities were required to set such effluent standards as they thought fit and, apart from the Royal Commission standard, most authorities decided their own standards. Today, for a very wide range of toxic, persistent or bioaccumulative substances there are mandatory national EQSs set by the DoE or by EC Directives. These EQSs apply to the receiving waters and are used to calculate effluent standards to ensure that the EQSs are not exceeded.

The EC Framework Directive on Dangerous Substances (1976) lists two groups of substances for priority action:

List I (the 'black' list) of very dangerous substances which must eventually be eliminated:

1. Organohalogen compounds.
2. Organophosphorus compounds.
3. Organotin compounds.
4. Carcinogenic, mutagenic and teratogenic substances.
5. Mercury and its compounds.
6. Cadmium and its compounds.
7. Persistent mineral oils and hydrocarbons.
8. Certain persistent synthetic substances.

The European Commission has identified 129 List I substances, and regulations setting EQSs or emission standards have been agreed for mercury and cadmium and 13 organic compounds.

List II (the 'grey' list) where preventive action is required to reduce the potential to pollute:

1. List I substances for which no standards have been set.
2. Heavy metals other than in List I.
3. Biocides not in List I.
4. Substances which produce taste and/or odour in groundwater or are likely to produce such substances and render water unfit for human consumption.
5. Toxic or persistent compounds of silicon.
6. Phosphorus and its inorganic compounds.
7. Non-persistent mineral oils and hydrocarbons.
8. Cyanides and fluorides.
9. Substances harmful to the oxygen balance, e.g. ammonia.

Member states are required to set their own EQSs or emission standards for List II substances.

In the UK, the DoE has published a considerable number of approved and provisional EQSs for List II substances and also a few List I and all the UK 'red list' substances. The red list was introduced in the UK following the 1987 North Sea Ministerial Conference for substances which are particularly toxic, persistent or bio-accumulative. The red list overlaps to a considerable extent with the EC List I and is now used for IPC purposes. At the 3rd North Sea Conference in 1990, member states approved a list of 'priority hazardous substances'. This list comprises 36 substances/groups of substances of which 23 are included in the UK red list.

The EQSs are based on toxicity studies which are carried out on contract from DoE by the Water Research Centre at Medmenham and subsequently approved by the DoE after widespread consultation. Provisional standards are set where full information concerning the ecotoxicology and environmental impact of pollutants is not yet available. In all cases a safety factor is incorporated in the EQS. The safety factors are not standardized because of the variability of toxicological data, but where data are sparse larger safety factors are used. Where environmental concentrations cannot be measured at levels below one-tenth of the proposed EQS, the standard must be applied by estimation or by modelling the effects of the discharge containing the substance.

Table 6 provides examples of EQSs which are approved and have been in use for some time. However, some of these have been reviewed and are due to be replaced by more stringent standards.

WATER QUALITY CLASSIFICATION

Chemical Classification

In England and Wales, river waters are classified on the basis of water quality into five categories:

Class 1a: Water of high quality suitable for abstraction for public supply, for game or high-class fisheries and of high amenity value.

Class 1b: Water of less high quality than Class 1a but usable for substantially the same purposes.

Class 2: Water suitable for public supply after advanced treatment; supporting coarse fisheries; moderate amenity value.

Class 3: Waters polluted to an extent that fish are only sporadically present; may be used for low-grade industrial abstraction purposes; considerable further potential for further use if cleaned up.

Class 4: Waters which are grossly polluted and which are likely to cause nuisance.

Scottish rivers, which are not greatly used for potable supply, are classed in four very similar categories with the difference that Class 1 is not subdivided.

The criteria which are used to define the various categories in both these systems are dissolved oxygen, BOD and ammonia, but the Environment Agency supplements the data in some river stretches with toxicity to fish and some criteria from the EC Surface Water Directive.

Biological Classification

The biological classification system which has been used for many years is based on assessments of the abundance and diversity of benthic river invertebrates, the results of which are used to calculate pollution indices. Two methods are currently in use, the Biological Monitoring Working Party (BMWP) score system and the Average Score Per Taxon (ASPT) score. Both methods assign a numerical score which is related to (i) the relative abundance of key species or groups of animals, (ii) the sensitivity of these species or groups to pollution, (iii) their 'reliability' as indicators of pollution, and (iv) an estimate of overall species diversity or the total number of species present. These systems differ from each other in sensitivity to the degree of pollution in several ways such as along the length of rivers, sampling effort, and in assessment of pollution at different seasons. The RCEP examined their respective merits and concluded that ASPT was easier to use and more reliable, differentiating, for example, between sites which have similar BMWP scores but different physico-chemical characteristics.

The RCEP recommended the use of ASPT but only as an interim method pending the development of the River Invertebrate Prediction and Classification System (RIVPACS). The problem with all previous biological classification systems, including BMWP and ASPT, is that they take no account of the natural physical and chemical characteristics of river systems, which are dependent on local factors including climate, topography, geology and land use.

RIVPACS was developed by the Institute of Freshwater Ecology (IFE) in close collaboration with the NRA and RPBs. By intensive surveys of some 440 selected sites on 80 rivers where pollution was absent or minimal, and by relating the macro-invertebrate populations to a range of natural features (distance from source, width, depth, substrate, slope, height above sea level and chemical characteristics), the IFE produced a data base which, in conjunction with a computer model, can predict (for a given site) the community structure of the fauna which should be present in the absence of pollution. The fauna which is predicted can then be compared with the actual fauna found at the site to produce an Ecological Quality Index which should be a true reflection of the biological effect of pollution at that site. The RCEP believes that this system can be refined and extended to include other non-invertebrate fauna and plant species which may respond to subtle effects of pollutants which are not detected by the invertebrate fauna. With further development, RIVPACS should become an indispensable tool with high confidence limits for its predictions.

QUALITY OF UK WATERS

The overall quality of rivers, estuaries and coastal waters in England and Wales is published at five-year intervals by the DoE, while similar five-year surveys are published for Scottish waters by the Scottish Office Environment Department.

Tables 7 and 8 illustrate the changes in water quality in terms of kilometres of river length for England and Wales between 1958 and 1980, and between 1980 and 1990 respectively. Changes of water quality in Scotland are shown in Table 9 for the period 1980 to 1990.

It is apparent from these figures that, while in England and Wales there was

Table 7. River water quality in England and Wales, 1958–80

Class	1958 km	1958 %	1970 km	1970 %	1975 km	1975 %	1980 km	1980 %
Unpolluted	24 950	72	28 500	74	28 810	75	28 810	75
Doubtful	5220	15	6270	17	6730	17	7110	18
Poor	2270	7	1940	5	1770	5	2000	5
Grossly polluted	2250	6	1700	4	1270	3	810	2
Total	34 690		38 400		38 590		38 740	

Source: DoE

Table 8. River water quality in England and Wales, 1980–90

Class	1980 km	1980 %	1985 km	1985 %	1990 km	1990 %
Good 1a	13 380	34	13 470	33	12 408	29
Good 1b	14 220	35	13 990	34	14 536	34
Fair 2	8670	21	9730	24	10 750	25
Poor 3	3260	8	3560	9	4022	9
Bad 4	640	2	650	2	662	2
X	–	–	–	–	39	–
Unclassified	–	–	–	–	17	–
Total	40 630		41 390		42 434	

Source: DoE

Table 9. River water quality in Scotland, 1980–90

Class	1980		1985		1990*	
	km	%	km	%	km	%
1	45 184	95.0	45 510	95.6	46 111	96.9
2	1981	4.2	1688	3.5	1177	2.5
3	256	0.5	266	0.6	233	0.5
4	162	0.3	131	0.3	70	0.1
Total	47 583		47 595		47 591	

*Excludes rivers on islands which were not included in previous surveys (3142 km Class 1; 4 km Class 2).
Source: SOEnD, 1990.

considerable improvement between 1958 and 1980, this was followed by a worsening of pollution in the next decade. By contrast, in Scotland there was a continuing record of improvement between 1980 and 1990. The reason for the better performance in Scotland is attributed to the separation of the regulatory function from the operation of urban wastewater treatment plants, while, south of the border, the regional water authorities were both 'polluter' and 'gamekeeper' from 1974 until the setting up of the NRA in 1989.

These classifications are based on chemical quality (BOD, dissolved oxygen and ammonia), but this is likely to change in the future. The NRA suggested a combined classification scheme using chemical data with a biological 'override'. However, the RCEP is not in favour of this and wants to see the introduction of a comprehensive biological classification based on RIVPACS. It also wants the chemical classification to include more parameters, such as nitrate, phosphate and pesticides.

Estuarine classification throughout the UK is based on biological, chemical and 'aesthetic' quality and commenced in 1990 (1980 in England and Wales). Results for 1990 are shown in Tables 10 and 11, the only differences being that in Scotland classification is by surface area whereas south of the border it is by length.

In Northern Ireland, there are no long-term records of river water quality or of estuarine quality. In 1990, the river quality survey showed that, of 1435 kilometres 72% were in Class 1a and 1b, 24% in Class 2, 4% in Class 3 and only 0.2% in Class 4. Also, out of 120 kilometres of estuaries, 83 per cent were in Class A (good) and 17% in Class C (poor). There were no lengths of estuaries in Class B or D.

Table 10. Estuary water quality in Scotland, 1990

RPB	Area (sq km) in class shown (% of total in each class)				Total
	A	B	C	D	
Clyde	56.8 (63.6)	30.0 (33.6)	2.5 (2.8)	0.0 (0.0)	89.3
Forth	28.6 (32.9)	38.5 (44.3)	19.6 (22.5)	0.3 (0.3)	87.0
Highland	220.6 (84.2)	32.4 (12.4)	9.0 (3.4)	0.0 (0.0)	262.0
North East	1.6 (31.4)	3.7 (72.6)	0.0 (0.0)	0.0 (0.0)	5.3
Solway	262.7 (100.0)	0.0 (0.0)	0.0 (0.0)	0.0 (0.0)	262.7
Tay	70.8 (59.5)	48.2 (40.5)	0.0 (0.0)	0.0 (0.0)	119.0
Tweed	0.0 (0.0)	0.0 (0.0)	0.0 (0.0)	0.0 (0.0)	0.0
Total	641.1 (77.7)	152.8 (18.5)	31.1 (3.8)	0.3 (0.0)	825.3

Sources: SOEnD, 1990.

Table 11. Estuary water quality in England and Wales, 1990

Region	Percentage of estuary length in each class						Total (km)
	Good	Fair	Good and fair	Poor	Bad	Poor and bad	
	A	B	A and B	C	D	C and D	
Anglian	79	14	93	7	0.4	7	514
Northumbria	34	38	73	17	10	27	135
North West	49	23	72	13	15	28	452
Severn-Trent	14	61	75	25	0	25	56
Southern	75	21	96	4	0	4	388
South West	92	8	100	0	0	0	350
Thames	45	55	100	0	0	0	112
Welsh	78	20	98	2	0	2	420
Wessex	45	51	96	4	0	4	120
Yorkshire	12	43	55	45	0	45	40
Shared estuaries:							
Humber	43	57	100	0	0	0	65
Severn	61	39	100	0	0	0	71

Source: NRA, 1990.

Table 12. Coastal water quality in Scotland, 1990

Inshore waters	Length (km) in class shown (% of total in each class)				
RPB	A	B	C	D	Total
Clyde	2037.8 (78.7)	228.5 (8.6)	325.8 (12.3)	11.7 (0.4)	2639.8
Forth	29.3 (15.6)	62.1 (33.0)	79.9 (42.4)	17.0 (9.0)	188.3
Highland	3378.9 (96.5)	47.7 (1.4)	68.4 (2.0)	5.0 (0.1)	3500.0
North East	47.1 (18.7)	118.4 (47.0)	68.3 (27.1)	18.3 (7.2)	252.1
Solway	249.6 (89.5)	25.1 (9.0)	4.2 (1.5)	0.1 (0.0)	279.0
Tay	48.9 (60.0)	21.0 (25.8)	11.6 (14.2)	0.0 (0.0)	81.5
Tweed	43.1 (98.0)	0.7 (1.6)	0.2 (0.4)	0.0 (0.0)	44.0
Total	5870.7 (84.1)	503.5 (7.2)	558.4 (8.0)	52.1 (0.7)	6984.7

Sources: SOEnD, 1990.

In 1990, the coastal waters of Scotland were classified on the basis of 'aesthetic', biological, chemical and microbiological quality (Table 12) but, so far, there has been no classification of coastal waters in England and Wales or Northern Ireland.

Under the EC Bathing Water Directive, over 450 bathing waters around the UK coast are routinely monitored 20 times during each bathing season. In 1991, 343 of the 453 waters complied with the mandatory standard of 2000 *Escherichia coli* per 100 ml (76% compliance). The compliance rates for the period 1988–1994 are shown in Fig. 12.

It is to be noted that the DoE assesses compliance in terms of the coliform standards. However, there is a requirement that, if there is a deterioration in water quality or if their presence is suspected, tests should also be carried out for *Salmonella* and enteroviruses. The requirement is that no salmonellae or enteroviruses should be present in 1 litre and 10 litres respectively. Tests on a proportion of samples taken during the season (usually two or three) suggest that, if these results are included, a much higher proportion of bathing waters would fail to comply.

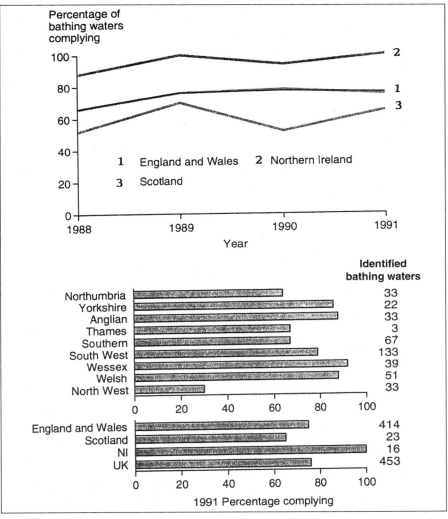

Fig. 12 Bathing waters: compliance with coliform standard by region and country, 1988 to 1991
(Source: Former water authorities, NRA, SOEnD, DOE(NI); reproduced by permission from HMSO)

5. FUTURE OUTLOOK

There is no doubt that the overall quality of rivers, estuarine and coastal waters has improved considerably in the UK since the early 1950s when a national system of river boards was set up and new powers were introduced to control pollution. Control of tidal waters by means of Tidal Waters Orders began in the 1960s. Very few were issued in England and Wales but, by 1970, about 40 Tidal Waters Orders, applying mainly to estuaries, had taken effect in Scotland. Only as recently as 1984 was the whole of the UK coast brought under control.

Despite the setbacks in river water quality in England and Wales between 1980 and 1990 referred to in the previous chapter, further improvements are now being carried out. The last decade has seen a large increase in environmental concern which has been reflected in environmental legislation in Europe and the UK. As evident at the 'Earth Summit' at Rio de Janeiro in 1992, environmental problems have become a global issue and there can be little doubt that, with the continuing rise of the 'green' movement, the pressure for higher standards will increase rather than decrease.

As an example of this pressure, the RCEP published its Sixteenth Report entitled 'Freshwater Quality' in 1992. This report was the most comprehensive yet produced by the Commission which spent six years in its preparation. The report took, as its guiding theme, the need for sustainability and made no fewer than 108 recommendations. It also emphasized the precautionary principle and stated *inter alia* that (i) there should be a general tightening of EQSs, (ii) there should be an incentive charging scheme with the aim of further reductions in pollutants, (iii) agriculture should become less dependent on pesticide usage, and (iv) in respect of eutrophication there should be a real attempt to restore rivers to, as near as possible, their original condition. However, it is apparent that few of these recommendations will be put into effect in the near future.

It is important to note here that the new Environment Agencies have been provided with new powers to ensure compliance with discharge standards. It is hoped that the use of Enforcement Orders will prove more speedy and more effective than reliance, as at present, on the threat of prosecution through the courts – a slow and somewhat cumbersome procedure. The powers of prosecution will, of course, be retained for serious or persistent breaches of the law.

Also in 1992, the Oslo and Paris Commissions announced an accord which is designed to bring about further reductions in the inputs of toxic, persistent and bio-accumulative substances with priority for organohalogens. More recently, in 1994, the US Environment Protection Agency, in the light of serious concern at the impact of organochlorines in the Great Lakes, promised that the Administration would develop a national strategy for substituting, reducing or prohibiting the use of chlorine or chlorinated compounds. The agency is particularly concerned by the growing evidence that chlorinated compounds are implicated in damage to the hormonal, reproductive, endocrinal, immune and nervous systems.

The need for global controls is supported by recent evidence that, despite the banning of organochlorine pesticides in the northern hemisphere, organochlorine compounds, after an initial substantial decrease, are still evident in northern Europe and Canada, and are most likely originating in developing countries where their use is still growing. There is evidence that organochlorine pesticides used on crops in the tropics are largely evaporated into the atmosphere and subsequently precipitated in rainfall over large areas of the north.

In conclusion, it is envisaged that, for all the foregoing reasons, standards will still continue to be tightened for many years ahead and that further man-made chemicals will have to be phased out in order to protect the environment, particularly the aquatic environment, which is a 'sink' for many pollutants. Increasingly, industry will have to turn away from 'end-of-pipe solutions' to clean technology in which steps are taken throughout the manufacturing process to eliminate pollution at source, minimizing waste and energy loss. Therefore, in future, it should not be necessary for the water industry to remove pesticides from surface and groundwaters – they should be prevented from getting there in the first place, as indicated by the RCEP. Equally, sewage works and industrial effluent treatment plants should not have to remove (as they do at present at great cost) increasing amounts of trace organic micropollutants, but steps should be taken to avoid their presence in wastewaters.

Hitherto, we have seen our inland waters and particularly our seas as disposal points for waste, with the emphasis on reducing loads to a level where 'self-purification' will take care of what remains. We are now beginning to recognize that this is no longer an acceptable policy and a radically new approach is required. Such a change in attitude will not take place overnight, but the signs are all around that this will happen in the near future. Having largely eliminated the gross pollution of our inland and coastal waters, we are entering a new phase whose targets may prove even harder and longer to achieve. In this endeavour, all who are concerned with the management of water quality cannot do better than keep in mind the following words of the great ecologist K. Curry-Lindahl in 1972:

> 'Without water, plants and animals cannot survive. All living organisms contain water and use it externally. Water is our most precious mineral. From the beginning of history it has been the key to civilization and development. It is the largest single factor in the growth of population. The available water supply is a boundary line beyond which no society or nation, agricultural or industrial, can go. Perhaps no greater conservation problem faces mankind than that of keeping the waters clean, and maintaining adequate and quantitatively useful supplies of this natural resource.'

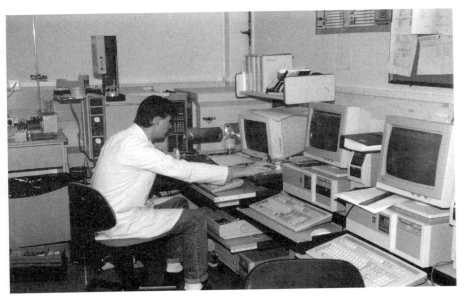

Plate 1. Gas chromatograph/mass spectrometer in use in laboratory

Plate 2. Use of electro-fishing apparatus to study fish populations

Plate 3. Collection of a sample of stream fauna by 'kick' sampling with a hand net

Plate 4. Sieving a grab sample of bottom sediment

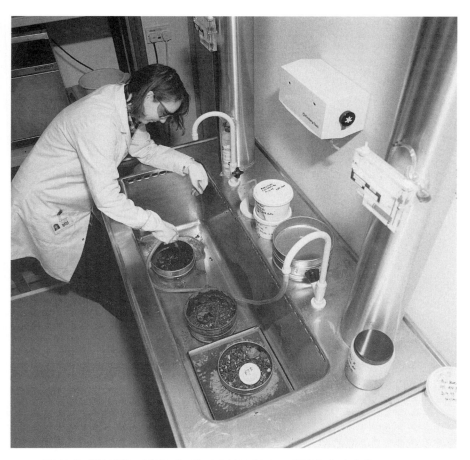

Plate 5. Washing biological samples in a ventilated sink (to remove formaldehyde vapour)

Plate 6. Laboratory examination of biological samples

Plate 7. Stream sampling for chemical analysis

Plate 8. Survey vessel for estuaries and coastal waters

Plate 9. Sampling distillery effluent in inshore waters

Plate 10. Reversing water bottle for collection of deep samples in coastal waters

Appendix

EUROPEAN COMMUNITY LEGISLATION

Since 1973, the European Community (now European Union) has adopted about 300 items of environmental legislation in the following eight sectors: water pollution, air pollution, chemicals, wildlife, noise, environmental assessment, information and finances, and international conventions. The following are the principal measures to control water pollution throughout the member states:

1976	Directive 76/160/EEC on the quality of bathing water. (Bathing Water Directive)
1976	Directive 76/464/EEC on pollution caused by certain dangerous substances discharged into the aquatic environment of the community. (Dangerous Substances Directive)*
1977	Decision 77/795/EEC on establishing a common procedure for the exchange of information on the quality of surface freshwater in the community. (Exchange of Information Decision)
1978	Directive 78/659/EEC on the quality of freshwater needing protection or improvement in order to support fish life. (Freshwater Fisheries Directive)
1979	Directive 79/923/EEC on the quality required of shellfish growing waters. (Shellfish-Growing Waters Directive)
1991	Directive 91/676/EEC on the pollution of waters caused by nitrates from agricultural sources. (Nitrates Directive)
1991	Directive 91/271/EEC on urban waste water treatment. (Urban Waste Water Treatment Directive)

*The Dangerous Substances Directive is a 'framework' Directive which has to date generated 10 'daughter' directives which regulate 13 chemical substances and their compounds or isomers.

BIBLIOGRAPHY AND FURTHER READING

Allen, J.D. 1995. *Stream Ecology*. Chapman and Hall. London.
Boon, P.J. and Howell, D.L. (Eds.) Freshwater quality: defining the indefinable. HMSO (Edinburgh). 1996.
Carlson, C.A. 1968. *Ecology*, **49**, 162–169.
Chapman, Deborah (ed.) 1992. *Water Quality Assessments: A Guide to the use of Biota, Sediments and Water in Environmental Monitoring*. Chapman and Hall for UNESCO/WHO/UNEP. London.
Department of the Environment, 1992. *The UK Environment*. HMSO. London
Ellis, K. V., 1989. *Surface Water Pollution and its Control*. Macmillan. London.
Hellawell, J.M. 1986. *Biological Indicators of Freshwater Pollution and Environmental Management*. Elsevier Applied Science Publishers.
Hynes, H.B.N. 1960. *The Biology of Polluted Waters*. Liverpool University Press.
Institution of Water and Environmental Management, 1991. *River Management*: Book No. 2*. IWEM. London.
Ibid., 1993. *Drinking Water Quality:* Booklet No. 3*. IWEM. London.
Ibid., 1994. *Water Supply in the UK:* Booklet No. 4*. IWEM. London.
Johnson, S.P. 1983. *The Pollution Control Policy of the European Communities*. London: Graham and Trotman.
Kirkwood, R.C. and Longley, A.J. (eds.) 1995. *Clean Technology and the Environment*. Glasgow: Blackie Academic and Professional.
Maitland, P.S. Boon, P.J. and McLusky, D.S. (eds.) 1994. *The Freshwaters of Scotland*. Wiley. Chichester.
McLoughlin, J. and Bellinger, E. G. 1993. *Environmental Pollution Control*. (International Environmental Law and Policy Series). Graham and Trotman/Martinus Nijhoff. London.
Mellanby, H. 1953. *Animal Life in Fresh Water*. Methuen. London.
Mellanby, K. (ed.), 1988. *Air Pollution, Acid Rain and the Environment*. Elsevier Applied Science Publishers. London.
Meybeck, M. Chapman, D. and Helmer, R. 1989. *Global Freshwater Quality: A First Assessment*. Blackwell Reference Publications. Oxford.
Moss, B. 1980. *Ecology of Freshwaters*. Blackwell Scientific Publications. Oxford.
National Rivers Authority 1990. *The Quality of Rivers, Canals and Estuaries in England and Wales*.
O'Riordan, T. (ed.) 1995. *Environmental Science for Environmental Management*. Longman Scientific and Technical.

*Copies of these publications are obtainable from the Institution's Headquarters.

Royal Commission on Environmental Pollution 1992. *Freshwater Quality:* Sixteenth Report. HMSO. London.

Russell-Hunter, W.D. 1970. *Aquatic Productivity: An Introduction to some Basic Aspects of Biological Oceanography and Limnology.* The Macmillan Company, Collier-Macmillan Ltd. London.

Ruttner, F. (trans. by Frey, D.G. and Fry, F.E.G.), 2nd Edition, 1953. *Fundamentals of Limnology.* University of Toronto Press.

Scottish Office Environment Department. *Water Quality Survey of Scotland, 1990.*

Solbe, J.F. de L.G. (ed.) 1986. *Effects of Land Use on Water Quality.* Ellis Horwood for the Water Research Centre.

Suess, M.J. (ed.) 1982. *Examination of Water for Pollution Control: A Reference Handbook.*
Vol. 1. *Sampling Data, Analysis and Laboratory Equipment.*
Vol. 2. *Physical, Chemical and Radiological Examination.*
Vol. 3. *Biological, Bacteriological and Virological Examination.*
Published on behalf of WHO Europe by Pergamon Press.

Sutcliffe, D.W. (ed.) 1994. *Water Quality and Stress Indicators in Marine and Freshwater Systems: Linking Levels of Organization.* The Freshwater Biological Association.

United Kingdom. 1995. *This Common Inheritance:* UK Annual Report on the UK's Sustainable Development Strategy of 1994 (including the Environmental Strategy of 1990). HMSO. London.

Whitton, B.A. 1975. *River Ecology:* Studies in Ecology, Vol. 2. Blackwell Scientific Publications.

World Health Organization. 1984. *Guidelines for Drinking Water Quality.*
Vol. 1. *Recommendations.*
Vol. 2. *Health Criteria and other supporting information.*
Geneva: WHO.

INDEX
(Italic page numbers denote figures or tables)

Accreditation	52
Acid deposition	38, *39*, 49
Acid rain	2–3, 16–17, 20, 24, 49
Acidification	17, 49
Acidity	16, 23, 38, 49
Acts of Parliament	
Rivers (Prevention of Pollution) Acts	53
Control of Pollution Act 1974	53
Water Act 1989	53
Environmental Protection Act 1990	54, 57
Water Resources Act 1991	57
Environment Act 1995	55
Aesthetic quality	64, 66
Agriculture	15, 22, 25, 55, 58, 68
Algae	
Blue-green (*see* Cyanobacteria)	
Brown	38
Diatoms	26, 37, 38
Green	37, 38
Red	38
Algal blooms	13, 25, 37, 38
Alkalinity	16, 23, 24, 38
Aluminium	24, 49
Amazon, River	11, 36
Ammonia	25, 45
Ammonification	25
Anaerobic decomposition	7
Analytical quality control	50
Apparent colour	10, 14
Aquatic communities	31–3
Aquatic fauna	35–6, 40–2, *41*, *43*, *44*, 62
Aquatic flora	35, 36–40
Aquifers	4, 12
Atomic absorption spectrometry	23, 26, 46

Atomic emission spectrometry	20, 22, 26, 46
Automated monitoring systems	52
Average score per taxon	62
Bacteria	
Denitrifying	25
Iron	36
Nitrifying	25
Pathogenic	37, 48
Sulphur	36
Baltic Sea	2
Bathing waters	48, 66, *67*
Beggiatoa alba	36
Bicarbonate	16, *17*, 23, 24
Bioaccumulation	46, 48, 56
Biochemical oxygen demand	45
Biological characteristics	30–42
Biological classification	62
Biological indicators	31
Biological Monitoring Working Party	62
Biological surveys	30–1, 33–5, 42
Biomagnification	46, 48
Biomass	33
Black-list substances	59–60
Blanket weed	40
Buffering	23–4
Calcium	22
Carbon dioxide	16, *17*, 22, 23, 29
Carbonate	16, *17*, 22, 23, 24
Carnivores	33, 40
Catchment management plans	58
Cellulose	37
Chalk	3, 12, 18, 23, 30, 49
Chemical characteristics	18–30
Chemical classification	61–2
Chitin	37

INDEX

Chloride 20, 24
Chlorinated compounds 69
Chlorophyll 33, 38
Cholera 48
Cladocera 42, *43*
Cladophera glomerata 40
Clarity 10
Classification of water bodies 11–12
Clyde, River 12, *44*, 50
Coastal waters 5, 7, 11, 12, 24, 28, 34–5, 38, 42, 50, 56, 66
Coliform standards 66
Colour 10, 14
Conductivity 16
Consent system 53, 55–7
Cooling water 32
Copepods 42
Critical loads 49
Crustacea 42
Curry-Lindahl, K. 70
Cyanobacteria 37, 38

Density 5–8, *6*
Detergents 45
Dissolved gases 18, *22*, 27–9
Dissolved inorganics 19–20, *21*, *22*
Dissolved organic carbon 27
Dissolved organics 27
Dissolved oxygen 28–9, *28*, *44*
Dissolved solids 15, 19–26, *21*
Diurnal variations 5, 13, 16, 23, 28–9, 40

Earth summit 54, 68
EC Directives 53, 54, 59, Appendix
 Bathing Water Directive 37, 48, 66
 Dangerous Substances Directive 31, 59

Urban Waste Water Treatment Directive 38, 46
Echinoderms 42
Ecological quality index 62
Electro-fishing 34, Plate 2
Enforcement orders 69
England and Wales 50, 53, 54, 57, 61, 63–6
 Department of the Environment 59, 61, 63, 66
 Environment Agency 55, 58, 61, 68–9
 HM Inspectorate of Pollution 54, 55
 National Rivers Authority 50, 53, 55, 57, 58, 62, 64
 Regional Water Authorities 64
Enteroviruses 37, 48, 66
Environmental action programmes 54, 55
Environment quality standards 31, *32*, 33, 50, 56, 57, 59–61, 68
Epilimnion 7, *8*
Erosion 3, 9–10, 15, 18
Escherichia coli 48, 66
Estuaries 5, 7, 11, 12, 28, 34, 42, 50, 51, 56, 64, 65
Europhication 38, 46, 57, 68
Eutrophic waters 11, 28, 29, 42
Exchange of information decision 51

Fertilizers 22, 40, 58
Fish 15, 17, 20, 24, 28, 30–1, 33, 34, 36, 45, 46, 49
Fisheries 31, 34, 41, 57
Food-processing wastes 45
Food web 33, *35*
Freezing 5–6
Fungicides 46–8
Future outlook 68–70

INDEX

Gas chromatography 48, Plate 1
Geology 3, 18, 23, 49
Glaciers 2, *3*, 10
Global environmental monitoring system 51
Great Lakes 69
Grey-list substances 60
Groundwaters 2, *3*, *4*, 12, 53

Hardness 18, 22, 39–30
Harmonized monitoring scheme 50–1
Herbicides 46–8
Herbivores 33, 40
Higher plants 38–40
HM Industrial Pollution Inspectorate 54, 55
HM Inspectorate of Pollution 54, 55
Hydrographic surveys 56
Hydrological cycle 2, *3*
Hypolimnion 7, *8*

Igneous rocks 18, 26, 49
Indicator species 31, *41*
Inert solids 46
Insecticides 46–8
Institute of Freshwater Ecology 62
Integrated pollution control 54, 55, 57, 60
International agreements 54
Irrigation 20

Joint Monitoring Group 51

Kick sampling 40, Plate 3

Lake District 24
Lakes 5, 7, 11, 19, 28, 29, 38, 51–2
Legislation 53–8
Leptospira 48

Light penetration 9, 10, 11, 15, 46
Lignin 37
Lime 17, 22, 29
Limestone 18, 23, 30, 49
Liquid chromatography 48
List I substances 59–60
List II substances 60

Macrobenthic fauna 34, 40–2, *43*, *44*
Macrobenthic invertebrates 40–2, *41*, 62
Magnesium 22
Marine waters 19, 24
Mass spectrometry 48, Plate 1
Metals *22*, 26, 46
Micronutrients 26
Minerals 9, 18
Mine waters 9, 14, 15, 24
Modelling of water quality 51
Molluscicides 46
Molluscs 30, 42
Monitoring 30, 50–1, 57
Mosses 40
Mussels 31–3

National Rivers Authority 50, 53, 55, 57, 58, 62, 64
Nephelometer 14–15
Neuston 9
Nile, River 11, 15, 36
Nitrate 25
Nitrite 25
Nitrogen cycle 25, 36
Nitrogen oxides 25, 49
North Sea Ministerial Conferences 51, 54, 60
Northern Ireland 53, 55, 64, 66
 Coastal waters 66
 Department of Environment 55

87

INDEX

Environment and Heritage
  Agency 55
River quality survey 64
Nutrients 13, 38, 46

Oligotrophic waters 29, 42
Optical quality 10–11
Organic wastes 45–6, *47*, 48
Organochlorine pesticides 37, 69
Oslo Commission 51, 54, 69
Oxygen 28–9, *28*

Paris Commission 51, 54, 69
Particulate inorganics 18–19
Particulate matter 15
Particulate organic carbon 27
Particulate organics 26–7
Pathogens 37, 48
Permethrin 31
Pesticides 37, 46–8, 58, 68, 69
Petroleum 45
pH 16–18, 23–4, 28–9, 38, *39*
Phosphates 25–6
Phosphorus 25–6
Photosynthesis 11, 13, 16, 23, 25, 29
Physical characteristics 13–18
Phytoplankton 25, 27, 29, 34, 37, 38
Pollution *41*, 42–5
Polychaete worms 42
Potassium 22
Primary consumers 33, *34*, 40
Primary producers 33, *34*, 40
Priority hazardous substances 60
Protein 25
Protozoa 48
Pycnocline 7

Quality control standards
  BSEN ISO 9002 52

NAMAS M10 52

Rainwater 2, 9, 16, 19–20, *19*, 24, 49
Red-list substances 60
Red Sea 2
Regulatory authorities 50, 52, 53, 55
Reservoirs 7–8, 12, 23, 38
Residence times *3, 4*, 12
Respiration 16, 23, 25, 29, 48
Reversing thermometer 13
River invertebrate prediction and
  classification system 62, 64
Rivers 11, 20, *21, 47*, 51, 52,
  56, 59, 61–4
Rotifers *43*
Royal Commission on Environmental
  Pollution 57, 58, 62, 64, 68, 69
Royal Commission on Sewage
  Disposal 59
Royal Commission Standard 59

Salinity 2, 7, 28
Salmonella 37, 48, 66
Sampling 5, 14, 40, 50, Plates 4–10
Scotland 3, 7, 17, 18, 24, *39, 44*, 53,
  54, 55, 57, 61, 64, 65, 66, 68
HM Industrial Pollution Inspectorate
  54, 55
Islands Councils 55
River purification authorities
  50, 53, 54, 55, 57
River purification boards
  50, 55, 57, 62
Scottish Environment Protection
  Agency 55
Scottish Office Environment
  Department 63
Sea outfalls 42, 56

INDEX

Seasonal variations	5, 12, 13, 15, 28–9, 38, 40
Seawater	2, 10, 24, 28
Seaweeds	35, 38
Secchi disc	14
Secondary consumers	33, 40
Sedimentary rocks	18, 24, 26, 30
Self-purification	36, 69
Sewage	18, 45, 48
Sewage effluents	40, 59
Shigella	37
Silicon	26
Sodium	20
Soft waters	3, 18, 30
Solvent ability	9, 18
Statutory water quality objectives	57
Sulphate	24
Sulphur dioxide	16–17, 49, 52
Supersaturation	23, 28, 29
Surface tension	9
Suspended solids	15
Temperature	5, *6*, 13, 48
Thames, River	12, 50
Thermal pollution	48
Thermal stratification	5, 7, *8*, 13
Thermocline	7, *8*
Tidal Water Orders	
Toxicity	30–1, 46, 48, 56, 59–61
Trace elements	*22*, 26
Transparency	13–14
Trophic pyramid	33, *34*, 40
Tropical waters	36
True colour	10, 14
Turbidity	10, 11, 14
Typhoid	48
UK groundwaters	12
UK water quality	*63*, 63–7
UN Conference on Environment and Development	54
UN Economic Commission for Europe	49
Urban Waste Water Treatment Directive	38, 46
US Environment Protection Agency	69
Viruses	
Enteroviruses	37, 66
Hepatitis	48
Polio	48
Viscosity	8
Visual quality	10, 57
Wales	3, 17, 18, 24, 50, 53, 54
Waste heat	48
Water cycle	2, *3*, *4*
Water properties	5–11
Water quality	4
assessment	50
classification	61–6
control	52
modelling	51
Water Research Centre	61
Water resources	4
Water supply	3
Weathering of rocks	9, 15, 18, 20, 23, 24, 26
Weil's disease	48
World Health Organisation	25
Zooplankton	10, 25, 27, 34, *43*

Printed by The Lavenham Press Ltd, Lavenham, Suffolk